Application of Gray System Theory in Fishery Science

Xinjun Chen
Editor

Application of Gray System Theory in Fishery Science

Springer

中国农业出版社
CHINA AGRICULTURE PRESS

Editor
Xinjun Chen
College of Marine Sciences
Shanghai Ocean University
Lingang New City, Shanghai, China

ISBN 978-981-99-0634-5 ISBN 978-981-99-0635-2 (eBook)
https://doi.org/10.1007/978-981-99-0635-2

Jointly published with China Agriculture Press
The print edition is not for sale in China (Mainland). Customers from China (Mainland) please order the print book from: China Agriculture Press.

This Springer imprint is published by the registered company Springer Nature Singapore Pte Ltd.
The registered company address is: 152 Beach Road, #21-01/04 Gateway East, Singapore 189721, Singapore

Preface

Since the Gray system theory was founded in 1982 by Professor Deng Julong, a renowned scholar in our country, its theory and methods have been continuously developing. At the same time, its application in different industries and disciplines has been deepening unceasingly, including fisheries science. As a result a series of good results were obtained, which created favorable conditions for the development of the Gray system theory. In 1998, the Shanghai Ocean University offered a graduate course entitled "Lecture on Gray Systems" for its postgraduate students. In 2003, the first textbook *The application of Gray system in fishery science* was compiled and published by China Agricultural Press. This book is a revised edition of *Application of Gray system in fishery science*. Based on the systematic introduction of the basic principles and methods of the Gray system, the book combines the research results of the Gray system in fishery science at home and abroad in recent years. The book is divided into eight chapters, covering the basic concept and theory of the gray system, original data processing and gray sequence generation, gray correlation analysis, gray clustering analysis, gray system modeling, gray prediction, gray decision-making, and gray linear programming.

This book is highly readable and practical. Its re-publication will offer new research methods and research tools for researchers engaged in fisheries science. The monograph can be used by scientific workers and research units engaged in fishery and marine biology; it is a good reference book and also can be used as teaching material for undergraduate and graduate students of fishery.

However, due to the limitations of length and reference materials, as well as the limited level of the authors, there may still be inappropriate points in this monograph. Therefore, readers are requested to make corrections and suggestions.

This book is supported by the Top fisheries disciplines of China, the high-level innovation team of local universities in Shanghai (the strategic innovation team of Oceanic Fishery Science and Technology), and the outstanding scientific research talents and innovation team of the Ministry of Agriculture (the sustainable development of oceanic squid resources).

Lingang New City, China Xinjun Chen

Contents

1 **Overview of Gray System Theory** 1
Xinjun Chen

2 **Raw Data Processing Method** 21
Xinjun Chen

3 **Gray Correlation Analysis** 35
Xinjun Chen

4 **Gray Cluster Analysis** 77
Xinjun Chen

5 **Basic Principles of Gray Dynamic Modeling** 105
Xinjun Chen

6 **Gray Prediction** 115
Xinjun Chen, Minyang Xie, and Jintao Wang

7 **Gray Decision** 165
Xinjun Chen

8 **Gray Linear Programming** 187
Xinjun Chen

Chapter 1
Overview of Gray System Theory

Xinjun Chen

Abstract Gray system theory is one part of the fields of control theory. It is the product of the viewpoint and method of cybernetics applied to social economic system and natural science system, and the combination of cybernetics and operational research. It takes the gray system as the research object, taking the whitening, desalination, quantification, modeling, and optimization of the gray system as the core and taking the prediction and control of the development of various gray systems as the goal. Gray system is between white system and black box, in which some information is known and some information is unknown. The gray system theory is aimed at the problem of uncertainty with little data and no experience, which is called "minority uncertainty." The sequence of systematic behavior is often irregular and varies randomly. For random variable and random process, people often use the method of probability and statistics. The method of probability statistics requires a large amount of data, so it is necessary to find statistical rules from a large amount of data. Gray system theory is not from the angle of looking for the statistical law and through a large number of samples to study, but with the method of number processing, will be chaotic original data collated into a more regular generating sequence. It is a kind of realistic law, not a priori law, to explore, discover, and seek the inner law from the disorderly original data. The main research contents of gray system are: the modeling theory of gray system, the relational analysis theory of gray factors, gray prediction theory and decision theory, gray system analysis and control theory, gray system optimization theory, and so on. In 1981, Professor Deng Julong, an expert of Chinese cybernetics, first put forward the concept of gray system. Since 1982, the gray system theory has been successfully applied in agriculture (including fishery), industry, meteorology, and other fields. In this chapter, the concept, characteristics, research contents, and development status of gray system theory are summarized, and the application of gray system in fishery is briefly introduced.

X. Chen (✉)
College of Marine Sciences, Shanghai Ocean University, Lingang New City, Shanghai, China
e-mail: xjchen@shou.edu.cn

X. Chen (ed.), *Application of Gray System Theory in Fishery Science*,
https://doi.org/10.1007/978-981-99-0635-2_1

Keywords Gray system theory · Gray correlation · Gray clustering · Gray modeling · Fishery science

The trend of highly integrated modern science and technology on the basis of a high degree of differentiation has led to the emergence of a group of disciplines of systems science with methodological significance. Systematic science reveals a deeper and more essential internal connection between things, which greatly promotes the holistic process of science and technology. Many complex problems that have long been difficult to solve in the field of science have been solved with the emergence of new disciplines of systems science, and people's understanding of the evolutionary laws of nature and objective things has gradually deepened due to the emergence of new disciplines of systems science. Systems theory, information theory, and cybernetics, which were born in the late 1940s, emerged from dissipative structure theory, synergetics, catastrophe theory, and fractal theory in the late 1960s and early 1970s, as well as the supercycles that appeared in the middle and late 1970s. Theories, dynamic system theory, and pansystems theory are all new disciplines of system science with horizontality and a cross-cutting nature (Chen 2003, 2023).

In systematic research, due to the existence of internal and external disturbances and the limitation of the level of understanding, the information obtained by people is often uncertain. With the development of science and technology and the progress of human society, people's understanding of the uncertainty of various types of systems has gradually deepened, and the study of uncertain systems has also become increasingly in depth. In the second half of the twentieth century, in the fields of systems science and systems engineering, various uncertain system theories and methods continuously emerged to form a large landscape. Examples include fuzzy mathematics created by Professor Lotfi Asker Zadeh in the 1960s (Syropoulos 2020), gray system theory created by Professor Deng Julong in the 1980s (Deng 1982), rough set theory created by Polish computer scientist Zdzisław I. Pawlak in the 1980s, and unascertained mathematics created by Professor Guangyuan Wang in the 1990s (Wang 1990). An emerging discipline in the study of uncertain systems. These new disciplines discussed the theories and methods for describing and processing various types of uncertain information from different angles and sides (Chen 2003, 2023).

1.1 Basic Concepts of the Gray System

In people's social economic activities, they often encounter incomplete or vague information. Many systems, such as society, economy, agriculture (fisheries), ocean, industry, and biology, have large uncertainties. For example, in fishery production, even if the information on operating fishing vessels, number of fishermen, operating hours, and operating fishing grounds is completely clear, due to unclear information such as fishery resources, marine environment, fishermen's technical level, sea conditions, market conditions, etc., it is difficult to accurately predict fishing yield, fishery output, and economic benefits. Similarly, due to changes in marine environmental conditions, it is difficult to accurately predict the location, trend, and yield of operating fishing grounds in fisheries forecasting.

The research object of gray system theory is the "poor information" uncertain system with "some information is already known, and some information is unknown." It achieves an accurate description and understanding of the real world through the generation and development of "partially known information."

1.1.1 Gray Meaning and Gray Phenomenon

In cybernetics, people often use the shade of color to describe the degree of clarity of information. For example, objects with unknown internal information are called black boxes. Under normal circumstances, we use "black" to indicate that the information is unknown, "white" to indicate that the information is completely clear, and "gray" to indicate that some information is clear and some information is not clear. The information is partially known and partially unknown, i.e., the information is incomplete, which is the basic meaning of "gray." In different situations, "gray" can be transformed or extended to different meanings. In nature and human society, the "gray" phenomenon is universal. The "gray" phenomenon refers to a phenomenon whose information is partially known and partially unknown. For example, a certain type of fishery resource is a gray phenomenon, and we can roughly estimate it. However, the amount of fishery resources cannot be accurately determined.

1.1.2 Gray System

The objective world is the material world and the world of information. However, in the fields of engineering technology, society, economy, agriculture, fishery, environment, ecology, and military, there are often situations of incomplete information. For example, the system factors or parameters are not completely clear, the relationship between the factors is not completely clear, and the system structure is not completely known. The mechanism of the system is not fully understood.

We call the system with completely clear information the white system. For example, in a circuit system, when the resistance value is given, there is a clear relationship between voltage and current, which is a white system, and it is a white system with a physical prototype.

A system with completely unclear information is called a black system. For example, a distant planet can also be regarded as a system. Although it is known to exist, it is completely unknown in terms of volume, mass, and distance from Earth. This is a black system.

A system with partially clear and partially unclear information is called a gray system. For example, in a fishery production system, fishery resources, water temperature, salinity, ocean currents, plankton, fishing vessels, fishing vessel parameters, fishermen, and fishery management measures are all factors that affect fishery yield. The mapping relationship between various factors and fishing yield is difficult to obtain. Obviously, the fishery production system is a gray system without a physical prototype.

1.2 Overview of the Development of Gray System Theory

1.2.1 The Scientific Background of Gray System Theory

According to the dialectical materialist view of science and technology, the emergence of any new science and theory has two aspects: inevitability and contingency. The law of the development of science and technology determines that in a certain historical period and at a certain development stage, new science and new theories will inevitably emerge. Gray system theory is also produced against a certain social development background. At the branch point of scientific development, Professor Deng Julong conformed to the needs of society and the law of scientific development and created gray system theory with great success.

Professor Deng has been engaged in the study of "prediction and control of economic systems" and "fuzzy systems" since the late 1960s and has been exposed to a large number of systems with some known and some unknown information. Use the method of fuzzy mathematics or probability theory to describe. Fuzzy mathematics mainly focuses on the phenomenon of "cognitive uncertainty" and uses the membership function to solve the problem based on experience. With no experience, no typical distribution conditions, and a small sample size, Professor Deng has conducted painstaking and fruitful research on this issue. Finally, in 1982, a new theory of gray systems was developed, and the first paper on gray systems was published in the journal *Systems and Control Letters* (Deng 1982). The editor-in-chief of the magazine, Professor Roger Brockett of Harvard University, commented on Professor Deng's first paper on gray systems: The term "gray system" was first created!

1.2.2 Overview of the Development of Gray System Theory

In 1982, the gray system theory established by the Chinese scholar Professor Deng Julong was a new method for studying the problem of uncertainty with little data and poor information (Deng 1982). Gray system theory uses "small sample" and "poor information" uncertain systems with "partially known information" and "poor information" as the research object. To achieve the correct description and effective monitoring of system operation behavior and evolution pattern. Many systems, such as society, economy, agriculture, industry, ecology, and biology, are named according to the field and scope of the research object, while the gray system is named according to the color.

The rapid development of gray system theory and its successful application in many scientific fields have won the recognition and attention of the international academic community. The English version of the international academic journal "The Journal of Gray System," founded in the UK in 1989, has become the core journal of important international abstract institutions such as the British Scientific Abstracts (SA) and the American Mathematical Review (MR). More than 300 academic journals around the world have accepted and published gray system papers. The Proceedings of the American Computer Society, Kybernetes, and other international journals published the gray system special edition. At present, many well-known scholars in the UK, the USA, Germany, Japan, Australia, Canada, Austria, Russia, and other international organizations are engaged in gray system research and have made important contributions to the development of gray system theory.

The application of gray system theory has been extended to many fields, such as industry, agriculture, society, economy, energy, transportation, petroleum, geology, water conservancy, meteorology, ecology, environment, medicine, education, sports, military, law, finance, etc. It solves a large number of practical problems in production, daily life, and scientific research. Since the 1980s, hundreds of cities, counties, and provincial regions in China have applied the methods, models, and techniques of gray system theory to study regional social, economic, and technological development issues, compile comprehensive development plans, and promote the health of the regional economy. Development: National, provincial, and municipal science funds actively support gray system research, and a large number of gray system theory or applied research projects are funded by various types of funds each year. According to statistics, more than 200 achievements in gray systems across the country have won national or provincial and ministerial awards. In 2002, scholars in gray systems in China won the World Organization for System and Control Award.

There are more than 100 universities in the world, such as Huazhong University of Science and Technology, Renmin University of China, Tsinghua University, Zhejiang University, Shandong University, Nanjing University of Aeronautics and Astronautics, University of Maryland, Toyohashi University, Kanagawa University, Vienna University of Economics, and France Aerospace Center, which have set up gray system theory courses. Huazhong University of Science and Technology and Nanjing University of Aeronautics and Astronautics in China recruit and train doctoral students in the field of gray systems.

Many publishing institutions at home and abroad, such as Science Press, National Defense Industry Press, Huazhong University of Science and Technology Press, Jiangsu Science and Technology Press, Shandong People's Publishing House, Science and Technology Literature Press and others, have published more than

60 academic books on the gray system. The Blue Book of Science and Technology in China (No. 8) compiled and published by the Ministry of Science and Technology of China affirmed gray system theory as a new soft science method created by Chinese scholars. At the same time, gray system theory has become a hot topic of attention and discussion in many important international conferences and will undoubtedly play a positive role in further understanding gray system theory in the world system science community.

In the past 40 years, gray system theory has established itself in the forest of science with its strong vitality, which has established its academic status as an emerging cross-disciplinary discipline (Liu et al. 2014). The vigorous vitality and broad development prospects of gray system theory are increasingly being recognized and valued by all walks of life at home and abroad. As an emerging discipline that is undergoing continuous development and improvement, gray system theory still has many problems that need to be further studied:

1. The connotation and exact description of the gray concept and the basic principles of the gray system;
2. The operation of the gray number gray algebraic system;
3. The information content of simple gray numbers and composite gray numbers;
4. The modeling mechanism, function, and application scope of different gray models;
5. The information and scientific basis for constructing the whitening weight function of the gray number;
6. The gray relational axiom, gray relational degree, and stability of relational order;
7. The construction, function, and qualitative and quantitative coupling points of practical buffer operators;
8. The properties of the gray nonnegative matrix, gray matrix spectral drift, and gray deepening research on input–output models;
9. Comparative research on uncertainty methods such as gray system theory, inexact set theory, unascertained mathematics, probability statistics, fuzzy mathematics, and innovation of uncertain mathematical theories;
10. Gray system theory application in various scientific fields and systems analysis, market forecasting, financial decision-making, asset evaluation, enterprise planning, and management decision-making at all levels of government.

1.3 Main Content of the Gray System Theory

After nearly 40 years of development, gray system theory has basically established a structural system of an emerging discipline. Its main contents include a theoretical system based on a gray hazy set, an analysis system based on gray correlation space, a method system based on gray sequence generation, and a model system with a gray model (GM) as the core. Evaluation, modeling, prediction, decision-making, control, and optimization are the main technical systems. Gray hazy sets, gray algebraic systems, gray equations, and gray matrices are the basis of gray system theory. Starting from the beauty and perfection of the disciplinary system, there are many issues worthy of further study. In addition to gray relational analysis, gray system analysis also includes gray clustering and gray statistical evaluation. Gray sequence generation is achieved through the function of the sequence operator. The sequence operator mainly includes the buffer operator (weakening operator and strengthening operator), the mean value generator, the ratio generator, the cumulative generator, and the accumulative generator. The gray model is constructed according to the five-step modeling idea. It weakens the randomness through the role of gray generation or the sequence operator and mines the potential pattern. Through the exchange between the gray difference equation and the gray differential equation, the discrete data sequence is used to establish the continuous data. A new leap in dynamic differential equations. Gray prediction is a quantitative prediction based on the GM model. According to its function and characteristics, it can be divided into several types, such as series prediction, interval prediction, catastrophe prediction, seasonal catastrophe prediction, waveform prediction, and system prediction. Gray decision-making includes gray target decision-making, gray relational decision-making, gray statistics, clustering decision-making, gray situation decision-making, and gray hierarchical decision-making. The main content of gray control includes the control problem of the intrinsic gray system and control based on the gray system method, such as gray relational control and GM (1, 1) predictive control. Gray optimization techniques include gray linear programming, gray nonlinear programming, gray integer programming, and gray dynamic programming.

Gray system theory mainly studies the "small sample uncertainty problem,", which is significantly different from the probability statistics of the "large sample uncertainty problem" and the fuzzy mathematics of the "cognitive uncertainty problem." Probability statistics, fuzzy mathematics, and gray system theory are the three most commonly used methods for studying uncertain systems. The study subjects all have a certain degree of uncertainty, which is the common point of the three. According to the results of Professor Deng's research, there are significant differences among the three (Table 1.1).

Fuzzy mathematics focuses on the problem of "cognitive uncertainty," and its research object has the characteristics of "clear connotation and unclear extension." For example, "young people" is a vague concept because everyone is very clear about the connotation of "young people." However, it is very difficult to delineate an exact range, in which the young people are within this range and the young people are not outside the range. The extension of the concept of young people is not clear. For this type of "cognitive uncertainty" problem with clear connotations and unclear extensions, fuzzy mathematics is mainly processed by the membership function based on the experience."

The study of probability statistics is the phenomenon of "random uncertainty," which focuses on the historical statistical law of the phenomenon of "random uncertainty" and examines the possibility of the occurrence of each of the "random

Table 1.1 Differences between gray systems, probability statistics, and fuzzy mathematics Chen (2003, 2023)

	Gray system	Probability statistics	Fuzzy mathematics
Connotation	Small sample size uncertainty	Large sample size	Cognitive uncertainty
Foundation	Gray hazy set	Kantoji	Fuzzy set
Basis	Information coverage	Probability distribution	Membership function
Means	Generate	Statistics	Boundary value
Characteristics	Little data	Multiple data	By experience
Requirement	Arbitrary distribution allowed	Typical distribution is required	Function
Goal	Law of reality	Historical statistics	Cognitive expression

uncertainty" phenomena with multiple possible outcomes. The starting point is a large sample, and the subject is required to obey a certain typical distribution.

Gray system theory focuses on the "small sample" and "poor information" uncertainties that are difficult to solve by probability statistics and fuzzy mathematics. It is characterized by "less data modeling." Different from fuzzy mathematics, gray system theory focuses on objects with "clear extension but unclear connotation." For example, by 2050, China will control the total population between 1.5 and 1.6 billion. This "between 1.5 and 1.6 billion" is a gray concept, and its extension is very clear. It is not clear which specific value is between 1.6 billion.

The main contents of gray system theory include the correlation analysis method of the mutual influence of factors; the gray statistics and gray clustering method based on the whitening weight function; the cumulative generation and the cumulative generation method for data processing; and the gray modeling for the establishment of a differential equation model. The methods include the gray prediction of series catastrophe prediction, seasonal catastrophe prediction, topological prediction, and system prediction; the gray decision-making method, gray linear programming, and gray dynamic control based on system behavior prediction.

1.3.1 Gray Relational Analysis

Gray relational analysis includes system factor analysis and system behavior analysis. Analysis of the factors that affect the main behavior of the system is called system factor analysis, while the quantitative comparison of the behavior of different systems is called system behavior analysis. For example, for the human–machine–environment system, the factors that affect the safety of the system include human physiological and psychological characteristics, operating skills, and health conditions. Environmental factors such as humidity, noise, and vibration. Among the above factors, it is necessary to analyze which factors are primary and which are secondary, which is the factor analysis of system security.

The gray correlation analysis method is a method to measure the degree of correlation according to the degree of similarity or difference between the factors of the system or the behavior of each system. Because the gray correlation analysis is based on the development trend, there is no excessive requirement on the size of the sample, and there is no need for the typical distribution pattern. The calculation is small. Even if there are more than ten variables, they can be calculated by hand, and there will be no grayscale. The quantitative results of the correlation are inconsistent with the qualitative analysis.

Probability statistics, fuzzy mathematics, and gray system theory are the three most commonly used methods for studying uncertain systems. The study subjects all have a certain degree of uncertainty, which is the common point of the three.

1.3.2 Gray Modeling

The mathematical model quantitatively expresses the mathematical relationship between system factors (variables). However, the general method can only be used to establish a difference equation model, which makes it difficult to analyze and describe the entire process of the system. Some scholars have pointed out, "For life science, economics, biomedicine, etc., we hope to establish differential equation models." Gray system theory has successfully solved the problem of differential equation modeling, which has always been considered difficult to solve due to the proposed new ideas and new methods, such as the generation number, the convergence of the discrete function, the limit and smoothness, the gray derivative, and the gray differential equation. The general formula of the dynamic differential equation model is GM (n, h), where n represents the order of the differential equation and the number of variables. Different values of n and h represent different system factor relationships, and their descriptive function is quite strong. Commonly used gray models are GM (1, 1), GM (2, 1), GM (0, 2), and GM (1, h).

Based on the premise of qualitative analysis and the backing of quantitative analysis, gray system theory proposes a five-step modeling method: a language model, network model, quantitative model, dynamic model, and optimization model.

It is worth noting that premodeling data generation and postmodeling residual identification are two unique and effective methods in gray theory. The accumulation of the original series is the most commonly used data generation method in gray modeling. This method provides intermediate information for modeling and weakens the randomness of the original data.

1.3.3 Gray Prediction

Gray prediction refers to the prediction made by the gray model GM (1, 1). According to its functions and characteristics, gray prediction can be divided into five categories: series prediction, catastrophe prediction, seasonal catastrophe prediction, topological prediction, and system prediction.

The prediction of the magnitude of the development and change in the system behavioral characteristics is called a series of predictions. The development and change of the system are continuous in time and orderly in space. Sequence prediction uses the time series or spatial sequence of the system to perform timing or fixed spatial prediction of the system. The collection of behavioral eigenvalues can be either equally spaced or nonequally spaced. In fact, sequence prediction studies the variation in behavioral characteristics over time or space.

Prediction of the abnormal value when the system behavior characteristic quantity will exceed a certain threshold is called catastrophe prediction. The feature of catastrophe prediction is to predict the time of occurrence of "catastrophe" or the occurrence time of anomalous numbers. The magnitude of the outlier is often a gray number with given upper and lower limits. For example, the forecast of a harvest year for a certain fishery resource is the forecast of the year when the annual average catch is relatively high (annual yield is more than 1000 tons), which is called the forecast of the harvest year, while the forecast of the poor year is that the annual average catch is too small (e.g., less than 400 tons). The prediction of the occurrence of a catastrophe in a certain season or a certain time zone of the year is called the prediction of seasonal catastrophe. For example, the forecast of fishing season or fishing season is the forecast of the occurrence of fishing in a specific time zone.

Topological prediction is the prediction of the characteristic data waveform of the system behavior over a period of time. Because many points can form a waveform, topological prediction specifies many given values. For each given value, a set of point distribution data can be obtained on the given curve, and then GM (1, 1) is established for each set of point distributions. The model predicts the time interval for the future development and change of this set of given values.

Predicting the relationship between several variables (factors) included in the system together and predicting the role of the dominant factors in the system is called system prediction.

1.3.4 Gray Decision-Making

The so-called decision-making refers to the selection of an appropriate countermeasure to address the occurrence of an event to achieve the best effect. Gray system theory proposes the gray decision because the establishment of the decision model is based on the gray class, that is, according to the gray model, especially the GM (1, 1) model.

Gray decision-making includes gray situation decision-making, gray-level decision-making, and gray planning. Gray situation decision-making refers to the decision made on the basis of the binary combination of events and countermeasures. Gray-level decision-making is a decision that coordinates, unifies, and compromises the intentions and functions of various decision-making groups (levels). Gray planning is a decision that combines general planning with gray prediction and gray number generation.

1.3.5 Gray Linear Programming

Linear programming is an important branch of operations research. It is a mathematical model that is widely used and easy to implement in the study of multivariable systems. It is also the most commonly used method for deterministic decision-making. The main problem it solves is how to maximize the role of limited resources (including human resources) and find effective ways to rationally use human, material, and financial resources. However, general linear programming has problems such as the inability to reflect the time-varying constraint conditions and gray parameters. If the gray system idea and modeling method are used to solve the problem, we will generally call it gray linear programming.

1.3.6 Gray Control

The execution of decisions is called control. The so-called gray control refers to the control of the intrinsic gray system, the control of the gray parameters in the system, or the predictive control composed of the GM (1, 1) model. The basic method of gray control is to find the pattern of system behavior development and change through the system behavior data series, predict the future behavior of the system according to the mastered pattern, and make control decisions based on the predicted value of future behavior.

Traditional control is control by judging whether the behavior that has occurred in the system meets the requirements, which is a kind of ex-post control. Its shortcomings are that it cannot be prevented in advance, it cannot be controlled in real time, and its adaptability is not strong. Gray predictive control is a kind of advanced control that can prevent problems before they occur, control them in a timely manner, and improve adaptability.

1.4 The Status and Characteristics of Gray System Theory in Scientific Development

1.4.1 The Position of Gray System Theory in the Discipline System

People have different understandings and perspectives on objective things, and the ways of dividing the disciplinary system are also different. In the seventeenth century, based on the understanding that scientific classification should correspond to human memory, imagination, and judgment, Bacon advocated dividing science into three categories: history, poetry and art, and philosophy. Later, Saint-Simon and Hegel proposed the idea of dividing disciplines according to metaphysics and idealism, respectively. In the late nineteenth century, Engels proposed dividing disciplines according to the different forms of material movement and their inherent order and establishing a scientific system structure, which laid a solid scientific foundation for the classification of disciplines.

In China, people usually divide science into natural science, social science, and thinking science according to different research objects, as well as philosophy and mathematics, which are summarized and run through the three fields. The basic disciplines of natural sciences are accustomed to being divided according to the six categories of mathematics, physics, chemistry, heaven, earth, and biology. Professor Qian Xuesen advocated that the entire science and technology system should be divided into six scientific fields: natural science, social science, systems science, thinking science, human science, and mathematical science. In terms of disciplinary division, we first classify scientific problems according to complexity and uncertainty and then point out the corresponding cross-disciplines with methodological significance according to the nature of various disciplinary problems, thus clarifying the cross-disciplinary group of gray system theory.

Use box (Ω) to represent the set of all things in the world. Circles A, B, C, and D are used to represent the set of simple things, complex things, deterministic things, and uncertain things, respectively, and the four-ring diagram of the classification of scientific problems can be obtained (Fig. 1.1). By marking the scientific methods for solving various problems, the four-ring diagram of the cross-disciplinary classification is obtained (Fig. 1.2).

By comparing Figs. 1.1 and 1.2, it can be seen that the grey system theory, as a scientific method to solve uncertain and semi-complex problems, achieves a new leap compared with pobability statistics and fuzzy mathematics to solve simple uncertain problems. However, the solution of complex and uncertain problems needs a new breakthrough in nonlinear science.

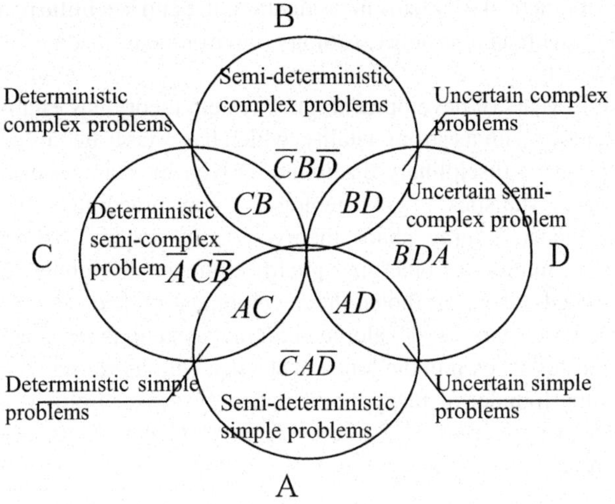

Fig. 1.1 Four-ring diagram of scientific problem classification (Chen 2003, 2023)

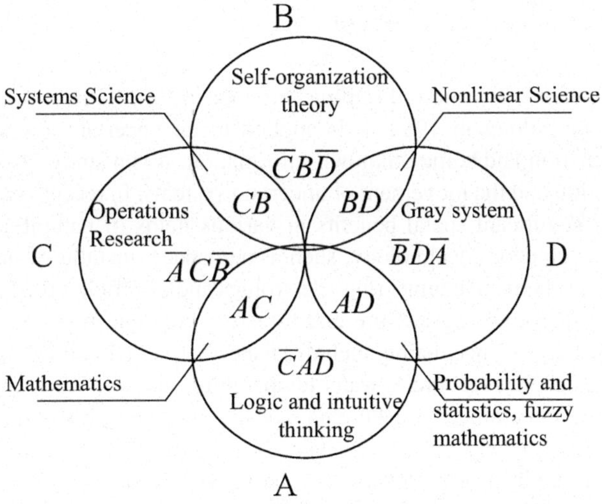

Fig. 1.2 Four-ring diagram of cross-disciplinary classification (Chen 2003, 2023)

1.4.2 The Role of Gray System Theory in the Discipline System

As a unique new theory, gray system theory has been recognized by the academic community at home and abroad and has played a huge role in the development of science. Its applications are widespread in agriculture, fisheries, industry, energy, transportation, petroleum, geology, and meteorology. It has successfully solved a large number of practical problems in production, life and scientific research in many scientific fields, such as hydrology, ecology, environment, medicine, military, economics, and society.

In the past 40 years of development, gray system theory has established itself in the forest of science with its strong vitality, which has established its academic status as an emerging cross-disciplinary discipline. Driven by gray system theory, "gray hydrology," "gray statistics," "gray geology," "gray breeding," "gray medicine," "gray control theory," "gray chaos theory," "gray system analysis of regional economy," and a number of emerging interdisciplinary disciplines have been successively produced, which promoted the development of science.

After nearly 40 years of development, gray system theory, as an emerging discipline, stands on its own in the forest of science with its strong vitality. Professor Xuesen Qian, the founder of fuzzy mathematics, Professor Lotfi A. Zadeh (USA), and the founder of synergetics, Professor Herman Haken (Germany), also spoke highly of the research on gray systems.

1.5 Research Progress on the Application of Gray System Theory in Fishery Science

Gray system theory is a new discipline founded by the famous Chinese scholar Professor Deng Julong in 1982. It is applied to an uncertain system with some known information and some unknown information. It is a study of small data and poor information and the movement of uncertain systems. In recent years, this theory has achieved significant social benefits in various fields of natural science, social science, and engineering technology, such as aerospace, metallurgy and petroleum, mechanical and chemical engineering, electronics and electricity, medical and health care, hydrometeorology, agriculture and forestry, and education and management. The field of fishery science mainly includes fishery economy and fishing production. Among them, fisheries resources, water temperature, salinity, ocean currents, plankton, fishing vessel parameters, fishermen, and fishery management measures are all factors that affect fishing yield. However, the fishery production system is a gray system without a physical prototype. The traditional probability statistical method, time series method, and linear regression analysis method require a large number of samples and follow a typical distribution. In the field of fishery science, where sample information is relatively scarce, the application of gray system theory can effectively solve many problems.

Based on the quantitative analysis of the China National Knowledge Infrastructure (CNKI) literature, the application of gray system theory in fishery science is divided into the following stages (Table 1.2): the initial stage (1988–1994), the middle stage (1995–2005), and the current stage (after 2006). The main research directions and progress of gray system theory in fishery science can be analyzed from the keywords of these three periods. From Table 1.2, the main research

Table 1.2 High-frequency keyword analysis results of the application of gray system theory in fishery science in different periods in China (Xie and Chen 2019)

Time	Keywords (frequency)
The initial stage (1988–1994)	Gray system theory (4); fishery yield (3); Prediction Model (1); yield prediction (1); gray correlation analysis (1); time series (1); Mariculture (1); marine fishery (1); Fishery production (1); marine fishing (1)
The middle stage (1995–2005)	Gray system theory (25); Prediction Model (17); marine fishing (11); gray correlation analysis (8); aquatic product yield (7); mariculture (6); Gray Clustering Method (6); fishing intensity (6); lake eutrophication (5); fishing yield (5); Marine Fisheries (5); fishery production (4); Comprehensive Evaluation (4); fishery resources (4); pond fish culture (4); lake water quality (3); yield relationship (3); fishery production (3); fishery economy (3); structural adjustment (3);
The current stage (after 2006)	Gray system theory (51); gray correlation analysis (49); Prediction Model (34); Marine Fisheries (25); fishery economy (24); Mariculture (23); marine fishing (22); aquatic product yield (20); aquatic product processing industry (18); influencing factors (17); Industrial Structure Adjustment (16); fishery production (13); fishery output (12); evaluation index (12); freshwater aquaculture (11); pelagic fisheries (9); time series (8); GM (1,1) Model (7); model precision (7); recreational fisheries (7)

direction in the early stage was the prediction of marine fishing or aquaculture production and gray correlation analysis; in the middle stage, on the basis of previous studies, the research on the aquaculture and marine fishing industry was strengthened, resulting in the environmental assessment of fishery waters. The current stage is the deepening of the fishery economic industry and the optimization of the gray forecasting model.

Gray system theory has been widely applied and theoretically studied in fishery science, mainly focusing on the following aspects: fishery economy, aquaculture, environmental assessment of fishery waters, and forecasting of fishing conditions.

1.5.1 Fishery Economy Industry

The fishery economy industry involves many components, and the methods of gray system theory applied are also diverse. It mainly includes industrial restructuring, evaluation of sustainable use of fishery resources, comparison of industrial competitiveness, analysis of factors affecting economic output, and regional economic division. The analysis of industrial structure adjustment is mainly to calculate the correlation between the total production of fisheries and the production of each part of the fishery through the gray correlation method and establish the GM (1, 1) model to predict the output of each part of the production, thereby making the proportion of each part of the industry structure. Recommendations for adjustment. For example, Song et al. (1999) and Song (2001) analyzed the correlation between the total fishing yield of Zhejiang Province and the yield of various operation methods, established the GM (1, 1) model for prediction, and obtained the correlation degree of various operation methods from 1980 to 1990. The order of size was fixed net, drift net, and trawl net, and trawl net, fixed net, and drift net in 1991–1997. In 2000, the total fishing yield of Zhejiang Province reached 33.5–36.5 million tons. Subsequently, the same analysis was conducted on the structural adjustment of the marine aquaculture industry in Zhejiang Province. In addition, many scholars have also used this method to analyze the industrial structure of fisheries and have achieved good results. Song et al. (1998) analyzed the current situation of marine fishing vessels in Shandong Province and used gray system theory to predict the fishing effort in 2000. Based on these results, they optimized the configuration of marine fishing vessels in Shandong Province. Song and Liu (2010) established the GM (1, 1) model using gray system theory to study the development trend of the number, total power, and average power of marine motor fishing vessels in China and found that the number of marine motor fishing vessels will show a downward trend. The total power and the average power showed an upward trend.

Since there are many uncertainties in the evaluation criteria for the sustainable use of fishery resources, gray system theory can be used to quantify them to evaluate the level of sustainable use, thereby understanding the development status of fishery resources. Chen (2001, 2003) analyzed the development of fishery resources in the East China Sea from 1978 to 1990 and selected a total of 23 indicators in the three evaluation index subsystems of the resource environment, society, and economy, and the optimal value of each sample point was formed into the mother. The index sequence of each sample point was used as a subsequence, and the different weights of each indicator were determined by the analytic hierarchy process. The correlation between each sample point and the parent sequence was calculated as the evaluation standard. The results showed the fishery resources in the East China Sea from 1978 to 1990. In particular, the sustainable utilization level was the lowest in 1983. Chen and Zhou (2004) proposed a comprehensive evaluation and evaluation model for the sustainable use of fishery resources based on gray system theory and combined it with the least squares criterion. This method has greater advantages than the traditional bioeconomic model and comprehensively reflects the model. Regarding various aspects of the sustainable use of fishery resources, the study used the development of fishery resources in the East China Sea from 1978 to 1990 as an empirical analysis. After 1978, the level of sustainable use of fishery resources in the East China Sea showed a downward trend. The lowest level of sustainable use in 1990 was only half of that in 1978.

1.5.2 Aquaculture Industry

The aquaculture industry is an important industry in fishery science, and its development trend can be used to measure the economic level of a country and affect the fishery economy and the income of fishermen. The application of gray system theory in the aquaculture industry mainly includes aquaculture yield prediction, aquaculture input–output ratio optimization, fishing port discharge forecasting, analysis of factors affecting the aquaculture industry, and selective breeding. Xie et al. (1998) used the five-step modeling method in gray dynamic control, i.e., the language model, network model, quantitative model, dynamic model, and optimization model, to analyze the investment and output value of grass carp and mud carp in the ponds of the Pearl River Delta. It is concluded that if 18% of the funds are extracted from the output of the system as the investment in the next year, the economic system can maintain a good state of development. Xie et al. (2000) used the gray correlation analysis method to analyze the relationship of multispecies polyculture fish in Shunde high-yield ponds and found that the stocking amount of carp had the greatest impact on the net yield of bighead carp, followed by grass carp stocking on the net yield of carp. The minimum is the relationship between silver carp stocking and the net yield of grass carp. The stocking of carp is the dominant parent factor, and the harvest of bighead carp is the dominant daughter factor. The gray dynamic model GM (1, 2) was used to establish the input–output relationship model for the main fish in the high-yield pond. Peng and Chen (2017) analyzed the influencing factors of the marine shellfish aquaculture industry in China's coastal provinces from 2003 to 2015 and concluded that the most influential factors were the production of marine shellfish aquaculture and the number of professional employees, and they also used the approach of approximating the ideal solution. According to the Technique for Order Preference by Similarity to Ideal Solution (TOPSIS), the ranking of each coastal province was obtained, and the results showed that Shandong, Liaoning, and Fujian were the top three. In addition, Liu et al. (2014) used the gray correlation analysis method to analyze and evaluate the correlation between the growth shape and body mass of Japanese flounder and found that the factors with the largest correlation were the overall length, body length, tail stalk height, and trunk length, which can be used as an important evaluation indicator for the cultivation of high-yielding Japanese flounder.

1.5.3 Environmental Assessment of Fishery Waters

The application of gray system theory in the evaluation of fishery waters has achieved good results. The environment of fishery waters, such as reservoirs, lakes, and rivers, is a gray system. Due to the incomplete information provided by the limited spatiotemporal monitoring data, the relationship between pollutants and the environment is uncertain. The clustering analysis of the gray whitening weight function in gray system theory satisfactorily solves the problem of fuzzy classification of water quality grade evaluation and the inability to quantitatively evaluate water quality grade. For example, Wang et al. (2006), Yang (1995), Xie (1997), and Li et al. (2011) successively performed gray cluster analysis on the eutrophication levels of Poyang Lake, Dongchang Lake, and Dianshan Lake. The types were clustered, and good results were obtained. Zhao et al. (2017) analyzed the relationship between the water quality indicators at the monitoring points and the water quality evaluation criteria using gray correlation analysis and determined the water quality evaluation grades at the monitoring points according to the degree of correlation and the weight of the indicators.

1.5.4 Fisheries Forecasting

Fishery situation forecasting is an important branch of the field of fishery science. It refers to the forecasting of various elements of fishery resources within a certain period of time and within a certain range of waters in the future. There are uncertainties in both, and the application of gray system theory can solve many problems. For example, Li and Chen (2007) analyzed the spatiotemporal distribution and abundance of offshore mackerel resources and selected the maximum value of the total yield of mackerel at each longitude (latitude) between 1998 and 2003 as the optimal vector, which was used as the parent sequence. The total yield composition vector for each longitude (latitude) in each year was used as the subsequence. Chen and Zheng (2007) used the gray correlation method to study the spatiotemporal changes in skipjack resources of the tuna purse seine fisheries in the western and central Pacific Ocean during the period 1990–2001 and found that in the 12 years 1998, 2000, 1994, 1995 and 1999, the abundance of skipjack resources in 1990, 1991, 1997, and 2001 was relatively low, and the abundance of skipjack resources in 1996, 1992, and 1993 was moderate. Xu et al. (2012) used gray correlation analysis to analyze the proportion of the optimal surface water temperature range (P_F) of the operating fishing grounds and spawning grounds in the open waters of Peru between 2003 and 2010 as well as the Nino1 + 2 SSTA and fishing effort. Studies have shown that the La Niña event will increase the proportion of the optimal surface water temperature area of the operating fishing ground and form a wide upwelling current, which is conducive to the feeding and growth of the fish and to the recruitment of jumbo flying squid. Gao et al. (2017), Duan et al. (2018), and Wang et al. (2019) used the gray correlation method to analyze the relationship between the resource index and environmental and climatic factors. The different GM (1, N) models and the GM (0, N) model are used to make relatively accurate predictions of resource abundance for neon flying squid, Peruvian anchovy, and Argentine flying squid.

References

Chen XJ (2001) Evaluation of sustainable utilization of marine fishery resources. Nanjing Agricultural University, Nanjing. (In Chinese)

Chen XJ (2003) Application of gray system theory in fishery science. China Agricultural Press, Beijing. (In Chinese)

Chen XJ (2023) Application of gray system theory in fishery science. China Agricultural Press, Beijing. (In Chinese)

Chen XJ, Zheng B (2007) Spatial and temporal distribution of skipjack resources in Tuna Purse Seine fishery in the western and central Pacific Ocean [J]. Oceanogr Res 25(2):13–22. (in Chinese)

Chen XJ, Zhou YQ (2004) Study on the synthesis assessment of sustainable use of fisheries resources based on gray theory. J Fish Sci China 11(z1):91–95. (In Chinese)

Chen XJ, Tian SQ, Ye XC (2002) Study on population structure of flying squid in Northwestern Pacific based on gray system theory. J Shanghai Fish Univ 11(4):335–341. (In Chinese)

Deng JL (1982) Control problems of gray systems. Syst Control Lett 1(5):288–294

Duan DY, Chen P, Chen XJ et al (2018) The construction of biomass forecasting model for the anchoveta (*Engraulis ringens*) by the gray system model. J Shanghai Ocean Univ 27(2): 284–290. (In Chinese)

Gao X, Chen XJ, Yu W (2017) Forecasting model of the abundance index of winter-spring cohort of neon flying squid (*Ommastrephes batramii*) in the Northwest Pacific Ocean based on gray system theory. Haiyang Xuebao 39(6):55–61. (In Chinese)

Li G, Chen XJ (2007) Tempo-spatial characteristic analysis of the mackerel resource and its fishing ground in the East China Sea. Period Ocean Univ China 37(6):921–926. (In Chinese)

Li ZL, Ma QM, Xu SQ et al (2011) Application of gray clustering analysis in Dongchanghu. Trans Oceanol Limnol 3:139–144. (In Chinese)

Liu YX, Liu YJ, Zhou Q et al (2014) Gray relational analysis between main growth traits and body weight in Japanese flounder (*Paralichthys olivaceus*). J Fish Sci China 21(2):205–213. (In Chinese)

Peng DM, Chen PD (2017) Evaluation of Chinese marine shellfish aquaculture industry in coastal provinces based on gray relationship and TOPSIS. Chin Fish Econ 35(3):78–83. (In Chinese)

Song WH (2001) Application of gray system theory to structure adjustment of marine industry in Zhejiang province. J Zhejiang Ocean Univ 20(2):91–111. (In Chinese)

Song XF, Liu L (2010) Forecast for fishing vessels developing trend based on gray system theory. Fish Mod 37(1):56–59. (In Chinese)

Song XF, Qiu TX, Jiao ZG et al (1998) Linear optimization of marine fishing vessel distribution in Shandong province. J Fish Sci China 5(4):82–88. (In Chinese)

Song WH, Chi HF, Yang JH (1999) Application of gray system theory to ocean fishing structure adjustment in Zhejiang province. J Zhejiang Ocean Univ 18(4):296–300. (In Chinese)

Syropoulos A (2020) A modern introduction to fuzzy mathematics. Wiley, New York, NY

Wang GY (1990) Unascertained information and its mathematical treatment. J Haibin Archi Civ Eng Inst 23(4):1–9. (In Chinese)

Wang XC, Wang LQ, Peng ZR (2006) Eutrophic status and water quality grade evaluations of Lake Dianshan based on gray-clustering method. J Shanghai Fish Univ 15(4):497–502. (In Chinese)

Wang YF, Chen XJ, Chen P et al (2019) Prediction of abundance index of Argentine shortfin squid in the Southwest Atlantic Ocean based on gray system model. Haiyang Xuebao 41(4):64–73. (In Chinese)

Xie J (1997) Application of gray system theory in evaluation of lake eutrophication degree in China. J Hydrol 4:9–12. (In Chinese)

Xie MY, Chen XJ (2019) Advances in the application of bibliometrics-based gray system theory in fisheries science. Trans Oceanol Limnol 5:117–126. (In Chinese)

Xie J, Xiao XZ, Huang ZH et al (1998) The comparative study on factors analysis and yield model of high-yield fish-pond for the pearl river delta and Yangtze delta. J Shanghai Fish Univ 7(2): 102–106. (In Chinese)

Xie J, Huang ZH, Xiao XZ et al (2000) The relationships among polycultured fishes and an input–output dynamics model in high-yield fish ponds. Acta Ecol Sin 20(2):317–320. (In Chinese)
Xu B, Chen XJ, Li JH (2012) Preliminary study on the influence of water temperature on the recruitment of *Dosidicus gigas*. J Shanghai Ocean Univ 21(5):878–883. (In Chinese)
Yang H (1995) Application of gray clustering method in lake water eutrophication evaluation. Fish Modern 6:36–39. (In Chinese)
Zhao LM, Lu Q, Wang N et al (2017) Application of improved gray correlation analysis method in fishery water quality assessment. J North China Univ Sci Technol 39(2):110–114. (In Chinese)

Chapter 2
Raw Data Processing Method

Xinjun Chen

Abstract Data is the basic work of statistical analysis and modeling. The processing of raw data is very important in data analysis and modeling. Different raw data come from different sources and have different properties. The raw data usually include: (1) scientific experiment and observation data; (2) socioeconomic statistics; (3) production experience data; (4) decision-making and target data of relevant departments; (5) quantitative data of qualitative information, etc. These original data have the following four main characteristics: (1) different dimensions, (2) different magnitude, (3) most of the data have a certain randomness, (4) a large number of data have a certain degree of gray. Therefore, strictly speaking, the majority of the data collected are gray parameters, with varying degrees of gray. For most gray parameters, it is necessary to whiten or desalinate them in order to improve the whiteness and reduce the gray degree. Because of the above characteristics and problems of the original data, it is difficult and limited to build the mathematical model by statistical analysis, so the original data should be transformed according to the classification of the mathematical model. The main purposes of the transformation are: (1) to make the index data as normal distribution as possible; (2) to unify the dimensionality of the variables; (3) to transform the nonlinear relation of the two variables into linear relation; (4) to replace a group of original variable indexes with a new group of independent variables with a small number of indexes. The commonly used transformation methods are standardization transformation, range transformation, mean transformation, initial transformation, modularization transformation, moving average transformation, weakening operator, and strengthening operator transformation. In this chapter, we will focus on introducing the source of the original data and its characteristics and providing several methods of the original data transformation and use some examples to demonstrate.

Keywords Original data · Data transformation method · Case analysis

---

X. Chen (✉)
College of Marine Sciences, Shanghai Ocean University, Lingang New City, Shanghai, China
e-mail: xjchen@shou.edu.cn

X. Chen (ed.), *Application of Gray System Theory in Fishery Science*,
https://doi.org/10.1007/978-981-99-0635-2_2

2.1 Sources and Characteristics of the Original Data

The original data generally include the regional characteristics of natural resources, such as sea conditions, meteorology, hydrology, topography, landforms, animals and plants, and reflect the regional socioeconomic conditions and productivity levels, such as population and population density, fishing labor, sea area, and number of fishing vessels. The power of fishing vessels and the total fishery output value, fishing output value, and aquaculture output value. According to their nature, raw data can be roughly divided into (1) scientific experiment and observation data; (2) socioeconomic statistical data; (3) production experience data; (4) decision-making and target data of relevant departments; and (5) qualitative data and quantitative data.

Different data have different sources (Chen 2003, 2023). However, in summary, the main sources are (1) historical statistical data of national statistical departments and industry departments, which are mostly social and economic indicators, and (2) historical observation data and scientific experimental reports of relevant business departments, which are mostly natural. Factor indicators, such as the observation data of fishery resources and the environment in the ocean; (3) data obtained from typical field surveys by selecting representative units or years; (4) data accumulated by regional planning departments through collection, survey, observation, and calculation. (5) Data obtained by surveying and interviewing workers with practical experience, production and technical personnel, scientific research personnel, and management personnel; (6) decision-making data such as development plans and construction plans formulated by relevant national departments; and (7) data in other aspects. The various information and data obtained above are called raw data. These data sources are different, and their types are different.

From the perspective of utilization analysis, these data have the following main characteristics (Chen 2003, 2023):

1. Different dimensions. For example, the fishery output value is RMB, the fishery output is kg, the water temperature is °C, the operation time is days, the voyage is nautical miles, the fishing effort is ton, kilowatt, boat, number of people, and the catch per unit effort is ton/day, ton/hour, ton/kW, etc.
2. The order of magnitude is very different. Some numbers are only decimal, and some numbers are as large as hundreds of millions. For example, the output value of fisheries is calculated at hundreds of millions of yuan or 10,000 yuan, while labor productivity is only tens to hundreds of yuan; the amount of fisheries resources is tens of thousands or tens of thousands of tons.
3. Most of the data have a certain degree of randomness, especially the statistical or observed time series or occasionally measured values, whether it is natural indicators or economic data, all have random changes, and all have obvious swings.
4. A large amount of data has a certain degree of gray, and most of the data collected using the above methods are the average or statistical value of each sample point in the region, which is not an exact white parameter in time or space but rather an

exact white parameter in time or space. A gray number with upper and lower limits. For example, in a fishery resource and environmental survey conducted by a survey ship, the data obtained can only be the data value at a certain point at a certain time, but due to the limitations of conditions and instruments and equipment, the values will have errors, and the magnitude of this error value cannot be known, resulting in a gray zone. For example, the amount of precipitation in a certain area in a certain year is the average of each actual observation record in the area, and it is impossible to know due to the difference in the measurement method and the error caused by the time calculation. The same problem also exists in some economic statistics. Therefore, strictly speaking, most of the collected data are gray parameters with varying degrees of gray.

2.2 Several Methods of Raw Data Whitening and Initial Transformation

2.2.1 Collection and Whitening of Raw Data

For the vast majority of gray parameters, whitening or lightening is needed to increase the whiteness and reduce the grayness. That is, through the continuous supplementation of information, the gray parameter gradually becomes relatively close to the actual value. The main methods of data whitening are as follows (Chen 2003, 2023):

1. The multiyear average of the observation station closest to the sampling site was directly used. For example, the average value of seawater temperature and salinity can be used for many years or the average value of recent days.
2. According to the contour map of each factor, the interpolation method is used to calculate its value. For example, for surface water temperature, salinity, seafloor topography, and other indicators, each representative sample point does not have ready-made accurate observation values, which can be used in the "iso-temperature contour map," "iso-salinity map," and "topographic contour map." The whitening value was calculated using the interpolation method.
3. Determine reasonable data based on the actual situation and data characteristics. For example, the indicators that reflect the quality of the marine environment include organic matter, nitrogen content, and phosphorus content. When used, it is impossible to input all the factors into the model. The convenience of calculation can be attributed to a comprehensive index. That is, the latest data in the region are used to determine the various factor indicators of each sample point, and then the transformed data are obtained using the "range transformation" method, and a comprehensive number can be obtained by adding them. In this way, the problems of different dimensions and large differences in magnitude between various factors can be solved.

4. Prediction of economic indicators. The classification and zoning of the marine economy should have relative stability, and the applied comprehensive indicators should also reflect the dynamic characteristics of the marine economy. Therefore, when using the corresponding indicator data, it is necessary to first carry out the development forecast of the indicator data and then based on the forecast. The values are classified and divided. There are many prediction methods, and the gray system GM (1, 1) model and the exponential increasing rate model are commonly used.
5. Relative values with the same weight are used. According to the principles of statistics, each individual should have equal weights, but in fact, various statistical objects have unequal products and unequal shapes; that is, there is a problem of unequal weights. For example, the absolute values of indicators such as sea area, population yield, and output value at various sampling points are very different, and they are undoubtedly unequal weights. If we use relative numbers such as the proportion of sea area, population density, average yield per unit, and per capita yield within a specific range, we can solve the problem of unequal weight of the same factor at each sample point.

2.2.2 Several Commonly Used Data Transformation Methods

Due to the above characteristics of the original data, there are certain difficulties and limitations in establishing the mathematical model for statistical analysis. Therefore, it is necessary to transform the original data according to the type of mathematical model to be built. The purpose of the transformation is to (1) make the indicator data as normal as possible; (2) unify the dimensions between the variables; (3) transform the nonlinear relationship between the two variables into a linear relationship; and (4) use a new set of independent variables with a small number of indicators to replace a set of interrelated original variables (Chen 2003, 2023).

Different mathematical models have different requirements for indicator variables. Most multivariate statistical analyses require that the variables generally follow a multivariate normal distribution and have consistent dimensions. For example, discriminant analysis requires the variables to be normally distributed; regression analysis requires the dependent variables to be normally distributed and requires a close correlation between the respective variables and the dependent variables. Cluster analysis requires the dimensions of each variable to be consistent and independent of each other. Therefore, the data must be transformed in a targeted manner according to the requirements of the mathematical model.

The commonly used transformation methods mainly include the following (Chen 2003, 2023):

2.2.2.1 Standardized Transformation

The calculation formula is

$$X'_{ij} = \frac{X_{ij} - \overline{X}_j}{S_j}$$

($i = 1, 2, \ldots,$ the number of N samples; $j = 1, 2, \ldots,$ the number of P variables) where X'_{ij} is the transformed data; X_{ij} is the original data; $\overline{X}_j$ is the arithmetic mean of the jth variable, i.e., $\overline{X}_j = \frac{\sum_{i=1}^{N} X_{ij}}{N}$; S_j is the standard deviation of variable j, that is, $S_j = \sqrt{\frac{\sum_{i=1}^{N} (X_{ij} - \overline{X}_j)^2}{N-1}}$.

After the transformation, the average value of each variable is 0, and the variance is 1, showing a standard normal distribution. There is a unified dimension between the variables, and the degree of correlation between the two variables before and after the transformation is unchanged. In a geometric sense, the standardized transformation is equivalent to moving the coordinate origin to the position of the center of gravity (i.e., the average value). The standardized transformation is applicable to continuous data with different dimensions and different orders of magnitude.

The relevant data in the empirical analysis of the doctoral dissertation "Evaluation of the Sustainable Utilization of Marine Fishery Resources" by Prof. Chen (2001) from Shanghai Ocean University are used for illustration. The resource and environmental subsystems of the sustainable use system of fishery resources in the East China Sea from 1978 to 1984 are shown in Table 2.1.where X_1 is the trophic level of the catch, and the unit is level; X_2 is the proportion of the yield of high-quality fish in the marine catch, and the unit is %; X_3 is the proportion of the catch of nonselective fishing gear in the marine catch, X_4 is the average fishing yield per unit of motor fishing vessels, in the unit of ton/vessel; X_5 represents the average fishing yield per tonnage of motor fishing vessels, in the unit of ton/tonnage; and X_6 is the average fishing yield per unit of motorized fishing vessels and nonmotorized fishing vessels per kilowatt. The average fishing yield of the unit is ton/kilowatt.

Table 2.1 Data of the resource and environment subsystem of the sustainable use system of fishery resources in the East China Sea (Chen 2001)

Year	1978	1979	1980	1981	1982	1983	1984
X_1	2.64	2.72	2.73	2.72	2.64	2.63	2.54
X_2	63.19	59.12	46.48	51.06	48.18	38.6	41.03
X_3	43.60	41.10	56.90	58.50	62.20	64.50	67.70
X_4	69.79	59.45	51.05	43.16	36.68	29.15	24.84
X_5	2.61	2.24	1.55	1.48	1.44	1.30	1.26
X_6	1.18	1.05	1.04	0.96	0.94	0.88	0.89

In the resource and environment subsystems shown in Table 2.1, the units of each evaluation index are different and therefore need to be initialized. The mean values and standard deviations of the sequences X_1, X_2, X_3, X_4, X_5, and X_6 were calculated.

$$\overline{X}_1 = (X_{1978} + X_{1979} + \ldots + X_{1984})/7 = (2.64 + 2.72 + \ldots + 2.54)/7 = 2.66$$

$$\overline{X}_2 = (X_{1978} + X_{1979} + \ldots + X_{1984})/7 = (63.19 + 59.12 + \ldots + 41.03)/7 = 49.67$$

$$\overline{X}_3 = (X_{1978} + X_{1979} + \ldots + X_{1984})/7 = (43.6 + 41.1 + \ldots + 67.7)/7 = 56.36$$

$$\overline{X}_4 = (X_{1978} + X_{1979} + \cdots + X_{1984})/7 = (69.79 + 59.45 + \cdots + 24.84)/7 = 44.87$$

$$\overline{X}_5 = (X_{1978} + X_{1979} + \ldots + X_{1984})/7 = (2.61 + 2.44 + \ldots + 1.26)/7 = 1.70$$

$$\overline{X}_6 = (X_{1978} + X_{1979} + \ldots + X_{1984})/7 = (1.18 + 1.05 + \ldots + 0.89)/7 = 0.99$$

$$S_1 = \sqrt{\frac{\sum_{i=1}^{N} (X_{ij} - \overline{X}_j)^2}{N-1}} = \sqrt{\frac{(2.64 - 2.66)^2 + \ldots (2.54 - 2.66)^2}{7-1}} = 0.07$$

$$S_2 = \sqrt{\frac{\sum_{i=1}^{N} (X_{ij} - \overline{X}_j)^2}{N-1}} = \sqrt{\frac{(63.19 - 49.67)^2 + \ldots (41.03 - 49.67)^2}{7-1}} = 8.98$$

$$S_3 = \sqrt{\frac{\sum_{i=1}^{N} (X_{ij} - \overline{X}_j)^2}{N-1}} = \sqrt{\frac{(43.6 - 56.36)^2 + \ldots (67.7 - 56.36)^2}{7-1}} = 10.24$$

$$S_4 = \sqrt{\frac{\sum_{i=1}^{N} (X_{ij} - \overline{X}_j)^2}{N-1}} = \sqrt{\frac{(69.79 - 44.87)^2 + \ldots (24.84 - 44.87)^2}{7-1}} = 16.28$$

$$S_3 = \sqrt{\frac{\sum_{i=1}^{N} (X_{ij} - \overline{X}_j)^2}{N-1}} = \sqrt{\frac{(2.61 - 1.7)^2 + \ldots (1.26 - 1.7)^2}{7-1}} = 0.52$$

$$S_3 = \sqrt{\frac{\sum_{i=1}^{N} (X_{ij} - \overline{X}_j)^2}{N-1}} = \sqrt{\frac{(1.18 - 0.99)^2 + \ldots (0.89 - 0.99)^2}{7-1}} = 0.11$$

Then, the data are transformed into:

$$X'_{11} = \frac{X_{11} - \overline{X}_1}{S_1} = \frac{2.64 - 2.66}{0.07} = -0.29$$

Other data transformations are similar.

Table 2.2 Values of each indicator after transformation of the mean and standard deviation (Chen 2003, 2023)

Year	1978	1979	1980	1981	1982	1983	1984
X'_1	−0.29	0.86	1.00	0.86	−0.29	−0.43	−1.71
X'_2	1.51	1.05	−0.36	0.15	−0.17	−1.23	−0.96
X'_3	−1.25	−1.49	0.05	0.21	0.57	0.79	1.11
X'_4	1.53	0.90	0.38	−0.11	−0.50	−0.97	−1.23
X'_5	1.75	1.05	−0.29	−0.43	−0.51	−0.77	−0.85
X'_6	1.71	0.55	0.44	−0.31	−0.50	−1.05	−0.90

The sequences of the transformed resource and environment subsystems are shown in Table 2.2.

2.2.2.2 Range Transformation

The calculation formula is

$$X'_{ij} = \frac{X_{ij} - X_{j\min}}{X_{j\max} - X_{j\min}}$$

where X'_{ij} is the transformed data; Xij is the original data; $X_{j\max}$ is the maximum value of the original data of the jth variable; $X_{j\min}$ is the minimum value of the original data of the jth variable.

After range transformation, the data have a unified dimension, with a maximum value of 1 and a minimum value of 0, and all the data change between 0 and 1. The degree of correlation between the two variables before and after the transformation is unchanged, and its geometric meaning is equivalent to moving the coordinate origin to the minimum value. Range transformation is suitable for the transformation of continuous raw data with different dimensions and quantities.

The data in Table 2.1 were used for analysis, and the maximum and minimum values of each indicator were first obtained. They are

$X_{1\max} = 2.73$	$X_{1\min} = 2.54$
$X_{2\max} = 63.19$	$X_{2\min} = 38.60$
$X_{3\max} = 67.70$	$X_{3\min} = 41.10$
$X_{4\max} = 69.79$	$X_{4\min} = 24.84$
$X_{5\max} = 2.61$	$X_{5\min} = 1.26$
$X_{6\max} = 1.18$	$X_{6\min} = 0.88$

Then, the data are transformed into:

Table 2.3 Index values after range transformation (Chen 2003, 2023)

Year	1978	1979	1980	1981	1982	1983	1984
X'_1	0.53	0.95	1.00	0.95	0.53	0.47	0.00
X'_2	1.00	0.83	0.32	0.51	0.39	0.00	0.10
X'_3	0.09	0.00	0.59	0.65	0.79	0.88	1.00
X'_4	1.00	0.77	0.58	0.41	0.26	0.10	0.00
X'_5	1.00	0.73	0.21	0.16	0.13	0.03	0.00
X'_6	0.99	0.57	0.53	0.25	0.18	0	0.04

$$X'_{11} = \frac{X_{11} - X_{1\,\min}}{X_{1\,\max} - X_{1\,\min}} = \frac{2.64 - 2.54}{2.73 - 2.54} = 0.53$$

The rest of the data conversion is similar.

Then, the sequences of the resource and environment subsystems after the range transformation can be obtained, as shown in Table 2.3.

2.2.2.3 Averaging Transformation

The calculation formula is

$$X'_{ij} = \frac{X_{ij}}{\overline{X}_j}$$

where X'_{ij} is the transformed data; x_{ij} is the original data; $\overline{X}_j$ is the average of the jth variable.

The transformed data have a uniform dimension, with values greater than 0 and concentrated near 1. Its mathematical expectation value is 1, and the expectation value of the difference between the variable and the mean is 0. This transformation is applicable to proportional variables such as length, volume, and mass.

Using the data in Table 2.1 as an example for analysis, the average value of each series is obtained, and the corresponding transformation value is

$$X'_{11} - \frac{X_{11}}{\overline{X}_1} = \frac{2.64}{2.66} - 0.99$$

The rest of the data are similar.

The mean transformation sequence of the resource and environment subsystem can be obtained, as shown in Table 2.4.

2.2.2.4 Initialization Transformation

The calculation formula is

Table 2.4 Index values after mean transformation (Chen 2003, 2023)

Year	1978	1979	1980	1981	1982	1983	1984
X'_1	0.99	1.02	1.03	1.02	0.99	0.99	0.95
X'_2	1.27	1.19	0.94	1.03	0.97	0.78	0.83
X'_3	0.77	0.73	1.01	1.04	1.10	1.14	1.20
X'_4	1.56	1.32	1.14	0.96	0.82	0.65	0.55
X'_5	1.53	1.32	0.91	0.87	0.84	0.76	0.74
X'_6	1.19	1.06	1.05	0.97	0.94	0.88	0.90

Table 2.5 Index values after initial value transformation (Chen 2003, 2023)

Year	1978	1979	1980	1981	1982	1983	1984
X'_1	1.00	1.03	1.03	1.03	1.00	1.00	0.96
X'_2	1.00	0.94	0.74	0.81	0.76	0.61	0.65
X'_3	1.00	0.94	1.31	1.34	1.43	1.48	1.55
X'_4	1.00	0.85	0.73	0.62	0.53	0.42	0.36
X'_5	1.00	0.86	0.59	0.57	0.55	0.50	0.48
X'_6	1.00	0.89	0.88	0.81	0.79	0.74	0.76

$$X'_{ij} = \frac{X_{ij}}{X_{i1}}$$

where X'_{ij} is the transformed data; x_{ij} is the original data; X_{i1} is the initial value of the ith variable (the first data).

The data after the initial value transformation have a unified dimension, and each value is a multiple of the initial value, which is convenient for analyzing the correlation between the series of factors, so it is suitable for processing the statistical data of socioeconomic aspects.

The data in Table 2.1 are used as an example for analysis, and the above formula is used for initial value transformation:

$$X'_{11} = \frac{X_{11}}{X_{11}} = \frac{2.64}{2.64} = 1$$

$$X'_{12} = \frac{X_{12}}{X_{11}} = \frac{2.72}{2.64} = 1.03$$

$$\ldots$$

$$X'_{12} = \frac{X_{17}}{X_{11}} = \frac{2.54}{2.64} = 0.96$$

Other calculations are similar.

The sequence of the initial value transformation of the resource and environment subsystem can be obtained, as shown in Table 2.5.

2.2.2.5 Modular Transformation

The calculation formula is

$$X'_{ij} = \sum_{k=1}^{j} X_{ik}$$

where X'_{ij} is the transformed data; X_{ik} is the kth data of the jth variable.

This transformation accumulates the time data series once a year to form a new data series, i.e., generate a time series of numbers. This transformation can be used for time series forecasting. This is the modeling mechanism and method of gray system theory for establishing mathematical models, making predictions, and performing dynamic analysis.

The data in Table 2.1 are used as an example for analysis, and the above formula is used for modular processing:

$$X'_{11} = \sum_{k=1}^{1} X_{1k} = X_{11} = 2.64$$

$$X'_{12} = \sum_{k=1}^{2} X_{1k} = X_{11} + X_{12} = 2.64 + 2.72 = 5.36$$

$$X'_{13} = \sum_{k=1}^{3} X_{1k} = X_{11} + X_{12} + X_{13} = 2.64 + 2.72 + 2.73 = 8.09$$

$$\ldots$$

$$X'_{17} = \sum_{k=1}^{7} X_{1k} = X_{11} + X_{12} + \ldots + X_{17} = 2.64 + 2.72 + \ldots + 2.54 = 18.62$$

Other calculations are similar.

The sequence of the modularized transformation of the resource and environment subsystem can be obtained, as shown in Table 2.6.

Table 2.6 Indicator values after modular transformation (Chen 2003, 2023)

Year	1978	1979	1980	1981	1982	1983	1984
X'_1	2.64	5.36	8.09	10.81	13.45	16.08	18.62
X'_2	63.19	122.31	168.79	219.85	268.03	306.63	347.66
X'_3	43.6	84.70	141.60	200.10	262.30	326.80	394.50
X'_4	69.79	129.23	180.28	223.43	260.12	289.27	314.11
X'_5	2.61	4.85	6.40	7.88	9.31	10.61	11.87
X'_6	1.18	2.23	3.27	4.22	5.16	6.03	6.92

2.2.2.6 Moving Average Transformation

The calculation formula is

$$\overline{X}_i = \frac{X_{i-1} + X_i + X_{i+1}}{3}$$

To avoid numerical cycles, the following formula can also be used:

$$\overline{X}_i = \frac{X_{i-1} + 2X_i + X_{i+1}}{4}$$

$$\text{Or } \overline{X}_i = \frac{X_{i-2} + X_{i-1} + X_i + X_{i+1} + X_{i+2}}{5}$$

This transformation can weaken the randomness of time data, eliminate the errors in collecting statistical data to varying degrees, and improve the reliability and accuracy for further data processing.

The data in Table 2.1 are used as an example for analysis, and the above formula is used for moving average transformation processing:

$$X'_{11} = \frac{2X_{11} + X_{12}}{3} = \frac{2 \times 2.64 + 2.72}{3} = 2.67$$

$$X'_{12} = \frac{X_{11} + X_{12} + X_{13}}{3} = \frac{2.64 + 2.72 + 2.73}{3} = 2.70$$

$$X'_{13} = \frac{X_{12} + X_{13} + X_{14}}{3} = \frac{2.72 + 2.73 + 2.72}{3} = 2.72$$

$$\cdots$$

$$X'_{17} = \frac{X_{16} + 2 \times X_{17}}{3} = \frac{2.63 + 2 \times 2.54}{3} = 2.57$$

Other calculations are similar.

Then, the sequence of the resource and environment subsystem after sliding transformation can be obtained, as shown in Table 2.7.

Table 2.7 Index values after sliding transformation (Chen 2003, 2023)

Year	1978	1979	1980	1981	1982	1983	1984
X'_1	2.67	2.70	2.72	2.70	2.66	2.60	2.57
X'_2	61.83	56.26	52.22	48.57	45.95	42.60	40.22
X'_3	42.77	47.20	52.17	59.20	61.73	64.80	66.63
X'_4	66.34	60.09	51.22	43.63	36.33	30.22	26.27
X'_5	2.49	2.13	1.76	1.49	1.40	1.33	1.27
X'_6	1.14	1.09	1.01	0.98	0.92	0.90	0.89

2.2.2.7 Transformation of the Weakening Operator and Strengthening Operator

Let X be the original data sequence and D be the buffer operator. When X is the increasing sequence and the declining sequence, respectively:

1. If the buffer sequence XD has a slower growth rate (or decay rate) or a decrease in amplitude than the original sequence X, then the buffer operator D is called a weakening operator.
2. If the growth rate (or decay rate) of buffer sequence XD is faster or the amplitude increases compared to the original sequence X, then buffer operator D is called a strengthening operator.

2.2.2.7.1 Weakening Operator Transformation

Let the original data sequence $X = (x\,(1), x\,(2)..., x\,(n))$.
Let $XD = (x(1)d, x(2)d, \ldots, x(n)d)$

where $x(k)d = \frac{1}{n-k+1}[x(k) + x(k+1) + \cdots + x(n)]$; $k = 1, 2, \ldots, n$.

Then, when X is a monotonic increasing sequence, a monotonic decay sequence, or an oscillation sequence, D is the first-order weakening operator, and XD is the buffer sequence after the first-order weakening.
If $XD^2 = XDD = (x(1)d^2, x(2)d^2, \ldots, x(n)d^2)$,

where $x(k)d^2 = \frac{1}{n-k+1}[x(k)d + x(k+1)d + \cdots + x(n)d]$; $k = 1, 2, \ldots, n$.

Then, D^2 is the second-order weakening operator for the monotonic growth, monotonic decay or oscillation sequence, and XD^2 is the buffer sequence after the second-order weakening.

2.2.2.7.2 Enhanced Operator Transformation

Let the original sequence and its buffer sequence be $X = (x\,(1), x\,(2)..., x\,(n))$, and $XD = (x\,(1)\,d, x\,(2)\,d..., x\,(n)\,d)$, respectively.

where $x(k)d = \frac{x(1)+x(2)+\cdots+x(k-1)+kx(k)}{2k-1}$; $k = 1, 2, \ldots, n{-}1$, and $x(n)d = x(n)$.

Then, when X is a monotonically increasing sequence or a monotonically declining sequence, D is the first-order strengthening operator, and XD is the buffer sequence after the first-order strengthening.
If $XD^2 = XDD = (x(1)d^2, x(2)d^2, \ldots, x(n)d^2)$,

where $x\,(n)\,d^2 = x\,(n)\,d = x\,(n)$; $x(k)\,d^2 = \frac{x(1)d+x(2)d+\cdots+x(k-1)d+kx(k)d}{2k-1}$; $k = 1, 2, \ldots, n{-}1$.

Then, D^2 is a second-order strengthening operator for a monotonically increasing sequence or a monotonically declining sequence, and XD^2 is the buffer sequence after the second-order strengthening.

Taking the fishery output data of a certain district in Zhejiang Province from 1983 to 1986 as an example for analysis, $X = (10{,}155, 12{,}588, 23{,}480, 35{,}388)$, the unit is 100 yuan, and its growth rate of 51.6% is very strong every year from 1983 to 1986. The average annual growth rate reached 67.7%, especially from 1984 to 1986. However, due to the limitation of fishery resources and the limitation of fishing capacity and technology, its fishery output value cannot continue to grow indefinitely, and it is impossible to maintain such a high development speed in the future. If the existing numerical values are used to directly model the prediction, the prediction results are simply unacceptable. After analysis and discussion, it is believed that the high growth rate is mainly due to the low base, while the reason for the low base is the insufficient utilization of fishery resources in the past. Therefore, in the forecast of fishery output in the future, it is necessary to adopt the weakening sequence growth trend and introduce the second-order weakening operator. The specific calculation process is as follows:

$$x(1)d = \frac{1}{4-1+1} \times (10155 + 12588 + 23480 + 35388) = 20403$$

$$x(2)d = \frac{1}{4-2+1} \times (12588 + 23480 + 35388) = 23819$$

$$x(3)d = \frac{1}{4-3+1} \times (23480 + 35388) = 29434$$

$$x(4)d = \frac{1}{4-4+1} \times 35388 = 35388$$

Then, the first-order buffer sequence $XD = (20403, 23819, 29434, 35388)$.

$$x(1)d^2 = \frac{1}{4-1+1} \times (20403 + 23819 + 29434 + 35388) = 27261$$

$$x(2)d^2 = \frac{1}{4-2+1} \times (23819 + 29434 + 35388) = 29547$$

$$x(3)d^2 = \frac{1}{4-3+1} \times (29434 + 35388) = 32411$$

$$x(4)d^2 = \frac{1}{4-4+1} \times 35388 = 35388$$

Then, the second-order buffer sequence $XD^2 = (27261, 29547, 32411, 35388)$ is obtained. The GM (1, 1) model established using the second-order buffer sequence XD^2 shows that the average annual increase of 9.4% in the fishery output value from 1986 to 2000 is basically acceptable and consistent with the actual situation.

References

Chen XJ (2001) Sustainable utilization assessment of marine fisheries resources[D]. Nanjing Agricultural University, Nanjing. (In Chinese)

Chen XJ (2003) Application of Gray system theory in fishery science. China Agricultural Press. (In Chinese)

Chen XJ (2023) Application of Gray system theory in fishery science. China Agricultural Press. (In Chinese)

Chapter 3
Gray Correlation Analysis

Xinjun Chen

Abstract Correlation analysis is a simple and practical analysis technique, which is to find the association or correlation existing in a large number of data sets and describe the law and pattern that some attributes appear at the same time in a thing. There are many methods of correlation analysis, among which the gray correlation is one of the common methods. In the real natural system and social economy system, there are many gray systems, and there are many subsystems (factors) that affect the gray system; therefore, the gray system theory puts forward the concept of the gray correlation degree analysis to each subsystem and intends to analyze the gray correlation degree of each subsystem through certain methods, to seek the numerical relations among the subsystems (or factors) in the system, this is the basic concept of the gray correlation degree. Gray correlation analysis provides a quantitative measure for the development of a system, which is very suitable for dynamic process analysis. The concrete calculation steps of gray correlation analysis are as follows: (1) determining the reference series and the comparison series that influence the system behavior, (2) dealing with the reference series and the comparison series dimensionless, (3) calculating the gray correlation coefficient between the reference series and the comparison series, (4) calculating the degree of correlation, (5) ranking the degree of correlation. In this chapter, first, the concept and characteristics of gray correlation are introduced, second, several common calculation methods of gray correlation degree are described, it includes general calculation method, gray absolute correlation degree, gray relative correlation degree, and gray comprehensive correlation degree. Finally, the application of gray correlation in fishery science is analyzed. The application cases in fishery science mainly include (1) adjustment and analysis of industrial structure of fishery economy, (2) assessment of fishery resources, (3) evaluation of sustainable utilization of fishery resources, (4) evaluation of influencing factors of fishery water quality, (5) the application of aquaculture (such as food selection), and (6) the evaluation of fish growth model.

X. Chen (✉)
College of Marine Sciences, Shanghai Ocean University, Lingang New City, Shanghai, China
e-mail: xjchen@shou.edu.cn

X. Chen (ed.), *Application of Gray System Theory in Fishery Science*,
https://doi.org/10.1007/978-981-99-0635-2_3

Keywords Gray correlation · Gray relative correlation · Gray absolute correlation · Gray comprehensive correlation

3.1 The Concept of Gray Correlation and Its Application

There are many large and small systems in the objective world, which are composed of many factors. The relationship between these systems and the internal factors of the system is very complex. In particular, the randomness of changes in superficial phenomena tends to confuse people's intuition and obscure the essence of things, making it difficult for people to obtain sufficient and comprehensive information when understanding, analyzing, predicting, and making decisions, and it is difficult to form a clear concept. Therefore, we believe that the relationship between different systems is gray, and the relationship between various factors in the system is also gray. It is difficult to identify the main contradictions and find the main factors when it is not clear which factors are closely related and which factors are not closely related. Therefore, gray system theory proposes the concept of correlation analysis. The purpose is to analyze the main relationship between various factors in the system through a certain method, to identify the most important factors affecting the system and to grasp the main aspects of the contradiction. For example, which industry has the most significant impact on the composition of the total fishery output, thereby creating conditions for the healthy development of the fishery production system.

The measure of the correlation between two systems or two factors is called the degree of correlation. It describes the relative changes between factors in the development process of the system, that is, the relativity of indicators such as the magnitude, direction, and speed of changes. The basic idea of gray correlation analysis is to determine whether the relationship is close according to the degree of similarity of the geometric shapes of the sequence curves. If the relative changes of the two are basically the same in the process of system development, the closer the curves are, the greater the degree of correlation between the two is considered; otherwise, the degree of correlation between the two is smaller. Gray correlation analysis is a quantitative description and comparison of the development and change of a system. Only by clarifying the correlation between systems or factors we can have a more thorough understanding of the system and distinguish which are the dominant factors, which are the potential factors, which are the advantages, and which are the disadvantages. Therefore, when analyzing and studying a gray system, it is necessary to first determine how to find the correlation from the random time series and calculate the correlation degree to provide a basis for factor discrimination, advantage analysis, and prediction accuracy testing and lay a good foundation for system decision-making. Therefore, the correlation analysis between gray factors is essentially the basis of gray system analysis, prediction, and decision-making.

The correlation analysis of gray system theory is different from the correlation analysis of mathematical statistics, which is mainly manifested in the following aspects. First, their theoretical bases are different. Correlation analysis is based on the gray process of the gray system, while correlation analysis is based on the stochastic process of probability theory. Second, the analysis method is different. Correlation analysis is a comparison of time series between factors, while correlation analysis is a comparison of groups of factors. Third, the data volume requirements are different. Association analysis does not require too much data, while correlation analysis requires sufficient data. Fourth, the research focuses are different. Correlation analysis mainly studies dynamic processes, while correlation analysis mainly focuses on static studies. Therefore, the degree of relevance analysis is more adaptable and unique in the socioeconomic system.

In addition, regression analysis, analysis of variance, and principal component analysis in mathematical statistics are all methods for systematic analysis. These methods have the following shortcomings: (1) a large amount of data is needed, and it is difficult to find a statistical law with a small amount of data; (2) a sample is required to follow a typical probability distribution, and a linear relationship between the data of various factors and the data of the system characteristics is needed. The factors are unrelated to each other. This requirement is often difficult to meet. (3) There may be a phenomenon in which the quantitative results are inconsistent with the qualitative analysis, resulting in the distortion and inversion of the relationships and laws of the system. The gray relational analysis method makes up for the shortcomings caused by the use of mathematical statistical methods for systematic analysis. It is equally applicable to large sample sizes and whether there are patterns in the sample. Moreover, the calculation is very convenient, and there is no discrepancy between the quantitative results and the qualitative analysis results.

At present, the application of gray correlation analysis is very extensive, and it has almost penetrated into various fields of social and natural sciences, such as agriculture, fisheries, education, health, politics and law, environmental protection, military, geography, geology, petroleum, hydrology, meteorology, etc. In the field of social economy, good results have been achieved.

3.2 Several Calculation Methods of Gray Correlation

In this section, several commonly used methods for calculating the degree of gray correlation are introduced, including the general calculation method, the absolute degree of gray correlation, the degree of gray relative correlation, and the degree of comprehensive gray correlation (Deng 1987, 1990; Liu et al. 2014).

3.2.1 *General Calculation Method*

Now, assume that the system behavior sequence

$$X_0 = (x_0\ (1), x_0\ (2) \ldots, x_0\ (n))$$
$$X_1 = (x_1\ (1), x_1\ (2) \ldots, x_1\ (n))$$
$$\cdots\cdots$$
$$X_i = (x_i\ (1), x_i\ (2) \ldots, x_i\ (n))$$
$$\cdots\cdots$$
$$X_m = (x_m\ (1), x_m\ (2) \ldots, x_m\ (n))$$

where X_0 is the parent sequence, X_i is the subsequence, and $x_i\ (k)$ is the observation data of factor x_i at time k.

The calculation of the gray correlation degree generally includes the following steps: (1) transformation of the original data; (2) calculation of the difference sequence; (3) calculation of the maximum and minimum difference between the two poles; (4) calculation of the correlation coefficient; and (5) calculation of the gray correlation (Deng 1987, 1990; Liu et al. 2014). The details are as follows:

Step 1: Raw data transformation

Because the dimension (or unit) of each factor in the system is not necessarily the same, for example, the labor force is a person, the output value is 10,000 yuan, the output is tons, etc., sometimes the magnitude of the value is different, such as the per capita income of several hundred yuan and the grain yield per hectare. The cost is several thousand kilograms, the output value of some industries reaches tens of billions, and the output value of some industries is only tens of thousands of yuan. Such data are often difficult to directly compare, and their geometric curve ratios are also different. Therefore, it is necessary to eliminate the dimensions (or units) of the original data and convert them into a comparable data series. See Chap. 2 for the transformation and processing methods of the original data.

Taking the initial value transformation as an example, we have

$$X_i' = X_i / x_i(1) = \left(x_i'(1), x_i'(2), \cdots, x_i'(n)\right), I = 0, 1, 2, \ldots, m;$$

Step 2: Find the difference sequence. Remember

$$\Delta_i(k) = \left|x_0'(k) - x_i'(k)\right|$$
$$\Delta_i = (\Delta_i(1), \Delta_i(2), \cdots, \Delta_i(n))$$
$$k = 1, 2, \ldots, n; i = 0, 1, 2, \ldots, m;$$

Step 3: Find the maximum difference and minimum difference between the two poles. Remember

$$M = \max_i \max_k \Delta_i(k)$$

Table 3.1 Raw data (Chen 2003, 2023)

Serial number	1	2	3	4	5	6	7
X_0	2.64	2.72	2.73	2.72	2.64	2.63	2.54
X_1	63.19	59.12	46.48	51.06	48.18	38.6	41.03
X_2	43.60	41.10	56.90	58.50	62.20	64.50	67.70
X_3	69.79	59.45	51.05	43.16	36.68	29.15	24.84
X_4	2.61	2.24	1.55	1.48	1.44	1.30	1.26
X_5	1.18	1.05	1.04	0.96	0.94	0.88	0.89

$$M = \min_i \min_k \Delta_i(k)$$

Step 4: Find the correlation coefficient.

The parent sequence of the data transformation is $\{X_0(t)\}$, and the subsequence is $\{X_i(t)\}$. Then, the correlation coefficient $L_{0i}(k)$ between the parent sequence $\{X_0(k)\}$ and the subsequence $\{x_i(t)\}$ can be calculated using the following equation:

$$\gamma_{0i}(k) = \frac{m + \xi M}{\Delta_i(k) + \xi M}$$

where $\xi \in (0,\ 1)$ is the resolution coefficient, $k = 1, 2, \ldots, n$; $i = 1, 2, \ldots, m$.

The correlation coefficient reflects the closeness (closeness) of the two compared sequences at a certain moment. At $\Delta_{\min}$, $\gamma_{io} = 1$, and at $\Delta_{\max}$, the correlation coefficient is the minimum value. Therefore, the range of the correlation coefficient is $0 < \gamma \le 1$.

Step 5: Calculate the degree of gray correlation.

$$r_{0i} = \frac{1}{n}\sum_{k=1}^{n} \gamma_{0i}(k);\ k = 1,\ 2,\ \ldots,\ n;\ \mathrm{i} = 1,\ 2,\ \ldots,\ m;$$

If the weights at different moments are inconsistent, the gray relational degree can be defined as:

$$r_{0i} = \sum_{k-1}^{n} W_k\, r_{0i}(k)$$

$$k = 1, 2, \ldots; n_i = 1, 2, \ldots, m;$$

where $\sum_{k=1}^{n} W_k = 1$.

Assume that there are parent sequence X_0 and subsequences X_1, X_2, X_3, X_4 and X_5, and assume that the resolution coefficient is 0.5 and the weights are the same (Table 3.1).

Step 1: Find the initial value sequence

Table 3.2 Data after initialization (Chen 2003, 2023)

Serial number	1	2	3	4	5	6	7
X_0'	1.00	1.03	1.03	1.03	1.00	1.00	0.96
X_1'	1.00	0.94	0.74	0.81	0.76	0.61	0.65
X_2'	1.00	0.94	1.31	1.34	1.43	1.48	1.55
X_3'	1.00	0.85	0.73	0.62	0.53	0.42	0.36
X_4'	1.00	0.86	0.59	0.57	0.55	0.50	0.48
X_5'	1.00	0.89	0.88	0.81	0.79	0.74	0.76

Table 3.3 Values after difference series processing (Chen 2003, 2023)

Serial number	1	2	3	4	5	6	7
Δ_1	0.00	0.09	0.29	0.22	0.24	0.39	0.31
Δ_2	0.00	0.09	0.28	0.31	0.43	0.48	0.59
Δ_3	0.00	0.18	0.30	0.41	0.47	0.58	0.60
Δ_4	0.00	0.17	0.44	0.46	0.45	0.50	0.48
Δ_5	0.00	0.14	0.15	0.22	0.21	0.26	0.20

The initial value of each sequence is transformed, and the initial value is selected as the denominator for transformation. The transformed data is calculated by the following equation:

$$X_i' = X_i / x_i(1) = (x_i'(1), x_i'(2), \cdots, x_i'(7))$$
$$i = 0, 1, 2, \ldots, 5.$$

The transformed data are shown in Table 3.2.

Step 2: Find the difference sequence

By $\Delta_i(k) = \left|x_0'(k) - x_i'(k)\right|$, $i = 1, 2, \ldots, 5$, and we have Table 3.3.

Step 3: Find the difference between the two poles

$$M = \max_i \max_k \Delta_i(k) = 0.60$$

$$m = \min_i \min_k \Delta_i(k) = 0$$

Step 4: Calculate the correlation coefficient

Take $\xi = 0.5$,

$$\gamma_{0i}(k) = \frac{m + \xi M}{\Delta_i(k) + \xi M} = \frac{0.30}{\Delta_i(k) + 0.30}; \quad i = 1, 2, \ldots 5$$

Thus, Table 3.4 is obtained.

Step 5: Calculate the degree of gray correlation

Table 3.4 Correlation coefficient (Chen 2003, 2023)

Serial number	1	2	3	4	5	6	7
γ_1	1.00	0.76	0.50	0.57	0.56	0.44	0.49
γ_2	1.00	0.77	0.52	0.49	0.41	0.38	0.34
γ_3	1.00	0.63	0.50	0.42	0.39	0.34	0.33
γ_4	1.00	0.64	0.41	0.39	0.40	0.37	0.39
γ_5	1.00	0.68	0.67	0.58	0.59	0.54	0.60

$$r_{01} = \frac{1}{7}\sum_{k=1}^{7} \gamma_{01}(k) = 0.62$$

$$r_{02} = \frac{1}{7}\sum_{k=1}^{7} \gamma_{02}(k) = 0.56$$

$$r_{03} = \frac{1}{7}\sum_{k=1}^{7} \gamma_{03}(k) = 0.52$$

$$r_{04} = \frac{1}{7}\sum_{k=1}^{7} \gamma_{04}(k) = 0.51$$

$$r_{05} = \frac{1}{7}\sum_{k=1}^{7} \gamma_{05}(k) = 0.67$$

3.2.2 Absolute Degree of Gray Correlation

Assuming that the parent sequence $\{X_0\}$ and the subsequence $\{X_i\}$ have the same length, they are

$$X_0 = (x_0(1), x_0(2) \ldots, x_0(n))$$
$$X_i = (x_i(1), x_i(2) \ldots, x_i(n))$$

Then, the corresponding starting point annihilation sequence is

$$X_0^0 = (x_0(1), x_0(2), \ldots, x_0(n))$$
$$X_i^0 = (x_i(1), x_i(2), \ldots, x_i(n))$$

where $X_0^0(k) = x_0(k) - x_0(1)$

$$X_i^0(k)=x_i(k)-x_i(1)$$

Then, the formula for calculating the absolute gray correlation between X_0 and X_i is

$$\varepsilon_{0i}=\frac{1+|s_0|+|s_i|}{1+|s_0|+|s_i|+|s_i-s_0|}$$

where

$$|s_0|=\left|\sum_{k=2}^{n-1}x_0^0(k)+\frac{1}{2}x_0^0(n)\right|$$

$$|s_i|=\left|\sum_{k=2}^{n-1}x_i^0(k)+\frac{1}{2}x_i^0(n)\right|$$

$$|s_i-s_0|=\left|\sum_{k=2}^{n-1}\left(x_i^0(k)-x_0^0(k)\right)+\frac{1}{2}\left(x_i^0(n)-x_0^0(n)\right)\right|$$

The gray absolute correlation degree ε_{0i} has the following properties (Liu et al. 2014):

1. $0<\varepsilon_{0i}\leq 1$;
2. ε_{0i} is only related to the geometric shapes of X_0 and X_i and has nothing to do with their relative spatial positions; in other words, the translation does not change the magnitude of the absolute correlation degree;
3. Any two sequences are not absolutely unrelated, that is, ε_{0i} is always nonzero;
4. The greater the degree of geometric similarity between X_i and X_0, the greater ε_{0i};
5. When any observation data in X_0 change, ε_{0i} will change accordingly;
6. When the lengths of X_0 and X_i change, ε_{0i} also changes;
7. $\varepsilon_{0i}=\varepsilon_{i0}$.

Assume that there are parent sequence X_0 and subsequences X_1, X_2, X_3, X_4, and X_5 (Table 3.5).

Step 1: Zeroing the starting point

By $X_i^0(k)=x_i(k)-x_i(1)$ Available;

$$x_0^0(1)=x_0(1)-x_0(1)=0$$

$$x_0^0(2)=x_0(2)-x_0(1)=2.72-2.64=0.08$$

Similarly, other zeroing values can be obtained, as shown in Table 3.6.

Step 2: Seeking $|s_0|$, $|s_i|$ and $|s_i-s_0|$

Table 3.5 Raw data (Chen 2003, 2023)

Serial number	1	2	3	4	5	6	7
X_0	2.64	2.72	2.73	2.72	2.64	2.63	2.54
X_1	63.19	59.12	46.48	51.06	48.18	38.6	41.03
X_2	43.60	41.10	56.90	58.50	62.20	64.50	67.70
X_3	69.79	59.45	51.05	43.16	36.68	29.15	24.84
X_4	2.61	2.24	1.55	1.45	1.44	1.30	1.26
X_5	1.18	1.05	1.04	0.96	0.94	0.88	0.89

Table 3.6 Data with initial values after zeroing (Chen 2003, 2023)

Serial number	1	2	3	4	5	6	7
X_0^0	0.00	0.08	0.09	0.08	0.00	−0.01	−0.10
X_1^0	0.00	−4.07	−16.71	−12.13	−15.01	−24.59	−22.16
X_2^0	0.00	−2.50	13.30	14.90	18.60	20.90	24.10
X_3^0	0.00	−10.34	−18.74	−26.63	−33.10	−40.63	−44.95
X_4^0	0.00	−0.37	−1.06	−1.13	−1.17	−1.31	−1.35
X_5^0	0.00	−0.13	−0.14	−0.22	−0.24	−0.30	−0.29

$$|s_0| = \left| \sum_{k=2}^{6} x_0^0(k) + \frac{1}{2} x_0^0(7) \right| = 0.19$$

$$|s_1| = \left| \sum_{k=2}^{6} x_1^0(k) + \frac{1}{2} x_1^0(7) \right| = 83.59$$

$$|s_2| = \left| \sum_{k=2}^{6} x_2^0(k) + \frac{1}{2} x_2^0(7) \right| = 77.25$$

$$|s_3| = \left| \sum_{k=2}^{6} x_3^0(k) + \frac{1}{2} x_3^0(7) \right| = 151.92$$

$$|s_4| = \left| \sum_{k=2}^{6} x_4^0(k) + \frac{1}{2} x_4^0(7) \right| = 5.72$$

$$|s_5| = \left| \sum_{k=2}^{6} x_5^0(k) + \frac{1}{2} x_5^0(7) \right| = 1.18$$

$$|s_1 - s_0| = \left| \sum_{k=2}^{6} \left(x_1^0(k) - x_0^0(k)\right) + \frac{1}{2} \left(x_1^0(7) - x_0^0(7)\right) \right| = 83.78$$

$$|s_2 - s_0| = \left|\sum_{k=2}^{6}\left(x_2^0(k) - x_0^0(k)\right) + \frac{1}{2}\left(x_2^0(7) - x_0^0(7)\right)\right| = 77.06$$

$$|s_3 - s_0| = \left|\sum_{k=2}^{6}\left(x_3^0(k) - x_0^0(k)\right) + \frac{1}{2}\left(x_3^0(7) - x_0^0(7)\right)\right| = 152.11$$

$$|s_4 - s_0| = \left|\sum_{k=2}^{6}\left(x_4^0(k) - x_0^0(k)\right) + \frac{1}{2}\left(x_4^0(7) - x_0^0(7)\right)\right| = 5.91$$

$$|s_5 - s_0| = \left|\sum_{k=2}^{6}\left(x_5^0(k) - x_0^0(k)\right) + \frac{1}{2}\left(x_5^0(7) - x_0^0(7)\right)\right| = 1.37$$

Step 3: Find the absolute degree of association

$$\varepsilon_{01} = \frac{1 + |s_0| + |s_1|}{1 + |s_0| + |s_1| + |s_1 - s_0|} = \frac{1 + 0.19 + 83.59}{1 + 0.19 + 83.59 + 83.78} = 0.50$$

Similarly, we can obtain

$$\varepsilon_{02} = 0.50$$

$$\varepsilon_{03} = 0.50$$

$$\varepsilon_{04} = 0.54$$

$$\varepsilon_{05} = 0.63$$

3.2.3 *Gray Relative Degree of Relevance*

Assuming that the parent sequence $\{X_0\}$ and the subsequence $\{X_i\}$ have the same length and the initial value is not equal to zero, then their initial values are

$$X_i' = X_i / x_i(1)$$

$$X_0' = X_0 / x_0(1)$$

Then, the corresponding initial value sequence is

$$X_0' = \left(x_0^0(1), x_0^0(2), \cdots, x_0^0(n)\right)$$

$$X_i' = \left(x_i^0(1), x_i^0(2), \cdots, x_i^0(n)\right)$$

The gray relative degree of X_0 and X_i is

$$r_{0i} = \frac{1 + |s_0'| + |s_i'|}{1 + |s_0'| + |s_i'| + |s_i' - s_0'|}$$

where

$$|s_0'| = \left| \sum_{k=2}^{n-1} x_0'(k) + \frac{1}{2} x_0'(n) \right|$$

$$|s_i'| = \left| \sum_{k=2}^{n-1} x_i'(k) + \frac{1}{2} x_i'(n) \right|$$

$$|s_i' - s_0'| = \left| \sum_{k=2}^{n-1} \left(x_i'(k) - x_0'(k)\right) + \frac{1}{2}\left(x_i'(n) - x_0'(n)\right) \right|$$

The gray relative degree of X_0 and X_i is r_{0i}. The gray relative degree r_{0i} has the following properties (Liu et al. 2014):

1. $0 < r_{0i} \le 1$;
2. r_{0i} is only related to the rate of change of the sequence X_0 and X_i relative to the starting point and is not related to the size of each observation data.
3. The rate of change of any two sequences is not unrelated, i.e., r_{0i} is always nonzero;
4. The more consistent the rate of change of X_0 and X_i relative to the starting point is, the greater $\mathrm{r_{0i}}$ is;
5. If any observation data in X_0 or X_i are changed, r_{0i} will change accordingly; if the sequence length changes, r_{0i} will also change;
6. $r_{0i} = r_{i0}$.

Using the parent sequence X_0 and the subsequences X_1, X_2, X_3, X_4, and X_5 in the above example as the original data, the gray relative degree of correlation between the parent sequence and the individual subsequences is determined.

Step 1: Initialize the sequence.

From formula $X_i' = X_i / x_i(1)$, the initial value sequence is obtained, as shown in Table 3.7:

Step 2: Seeking $|s_0'|$, $|s_i'|$ and $|s_i' - s_0'|$

Table 3.7 Data after initialization (Chen 2003, 2023)

Serial number	1	2	3	4	5	6	7
X_0'	1.00	1.03	1.03	1.03	1.00	1.00	0.96
X_1'	1.00	0.94	0.74	0.81	0.76	0.61	0.65
X_2'	1.00	0.94	1.31	1.34	1.43	1.48	1.55
X_3'	1.00	0.85	0.73	0.62	0.53	0.42	0.36
X_4'	1.00	0.86	0.59	0.57	0.55	0.50	0.48
X_5'	1.00	0.89	0.88	0.81	0.79	0.74	0.76

$$|s_0'| = \left| \sum_{k=2}^{6} x_0'(k) + \frac{1}{2} x_0'(7) \right| = 5.57$$

$$|s_1'| = \left| \sum_{k=2}^{61} x_1'(k) + \frac{1}{2} x_1'(7) \right| = 4.18$$

$$|s_2'| = 7.27$$

$$|s_3'| = 3.32$$

$$|s_4'| = 3.31$$

$$|s_5'| = 4.50$$

$$|s_1' - s_0'| = \left| \sum_{k=2}^{6} (x_1'(k) - x_0'(k)) + \frac{1}{2} (x_1'(7) - x_0'(7)) \right| = 3.99$$

$$|s_2' - s_0'| = 7.08$$

$$|s_3' - s_0'| = 3.13$$

$$|s_4' - s_0'| = 3.12$$

$$|s_5' - s_0'| = 4.31$$

Step 3: Find the relative degree of gray correlation

$$r_{01} = \frac{1 + |s_0'| + |s_1'|}{1 + |s_0'| + |s_1'| + |s_1' - s_0'|} = \frac{1 + 5.57 + 4.18}{1 + 5.57 + 4.18 + 3.99} = 0.57$$

Similarly, we can obtain

$$r_{02} = 0.54$$

$$r_{03} = 0.59$$

$$r_{04} = 0.59$$

$$r_{05} = 0.57$$

3.2.4 Gray Comprehensive Correlation

Assuming that the parent sequence $\{X_0\}$ and the subsequence $\{X_i\}$ have the same length, and the initial values are not equal to zero, ε_{0i} and r_{0i} are the gray absolute and relative degrees of correlation between $\{X_0\}$ and $\{X_i\}$, respectively, $\theta \in [0, 1]$, then $\rho_{0i} = \theta\varepsilon_{0i} + (1 - \theta)r_{0i}$ is the gray comprehensive correlation between X_0 and X_i.

The gray comprehensive correlation degree not only reflects the degree of similarity of the polyline but also reflects the closeness of the change rate of X_0 and X_i relative to the starting point and is a quantitative indicator that more comprehensively describes whether the sequences are close. Generally, take $\theta = 0.5$.

The gray comprehensive correlation degree ρ_{0i} has the following properties (Liu et al. 2014):

1. $0 < \rho_{0i} \leq 1$;
2. ρ_{0i} is not only related to the size of each observation data of series X_0 and X_i but also related to the rate of change of each data relative to the starting point;
3. ρ_{0i} is always nonzero;
4. When the data in X_0 and X_i are changed, ρ_{0i} will also change accordingly;
5. When the sequence length of X_0 and X_i changes, ρ_{0i} also changes;
6. When θ takes different values, ρ_{0i} is also different;
7. $\rho_{0i} = \rho_{i0}$.

Using the above example to calculate the gray comprehensive relevance, we take $\theta = 0.5$; then, the corresponding comprehensive correlation degrees are

$$\rho_{01} = \theta\varepsilon_{01} + (1 - \theta)r_{01} = 0.5 \times 0.5 + (1 - 0.5) \times 0.59 = 0.545$$

Similarly, we obtain

$$\rho_{02} = 0.52$$

$$\rho_{03} = 0.55$$

$$\rho_{04} = 0.57$$

$$\rho_{04} = 0.60$$

3.3 Application Examples of Gray Relational Theory in Fishery Science

In this section, the application of the gray correlation method in fishery science and its case analysis are mainly described, and good research results have been obtained in the aspects of industrial structure analysis, resource fishing ground analysis, and fishery biology research. The main contents include (1) fishery economic industrial structure adjustment and analysis; (2) fishery resource assessment; (3) fishery resource sustainable utilization assessment; (4) influencing factor assessment in the water quality of fisheries; (5) application in the aquaculture industry; and (6) basic biological evaluation of fish growth models. Detailed analysis is now carried out based on relevant examples (Chen 2003, 2023).

3.3.1 Application of Gray Correlation in the Analysis of Fishery Industry Structure

In the paper "Analysis of Fishery Production Structure in China," Chen and Zhou (2002b) analyzed the fishery production structure in China in the past 50 years through the correlation method in gray theory and explored the change process of fishery production and its contribution to fishery development. Identify the problems in the development of China's fisheries and the factors that restrict development and provide a basis for decision-making for the sustainable development of China's fisheries.

The data are from the China Fisheries Statistical Yearbook (1949–1997). The data items include the total production of aquatic products, the production of seawater, the production of freshwater, the production of seawater fishing and aquaculture, the production of freshwater fishing and aquaculture, and the fish and shrimp of seawater and freshwater. Crabs, shellfish, algae, etc., and some major fishing and farming species. According to the development of China's fisheries, the analysis was conducted in three time periods: 1954–1977, 1978–1984, and 1985–1997. In this study, the general gray correlation degree calculation method was used, and the resolution coefficient was set to 0.5. The main analysis results are as follows:

3.3.1.1 Correlation Analysis of the Total Aquatic Production with the Production of Seawater and Freshwater

3.3.1.1.1 Correlation Analysis of the Total Aquatic Production and the Production of Seawater and Freshwater Between 1954 and 1977

Using the total production of aquatic products as the mother series, the correlation between the total production of aquatic products (X_0), the production of seawater (X_s), and the production of freshwater (X_f) from 1954 to 1977 was analyzed (Table 3.8).

The correlations between the total production of aquatic products and the production of seawater and freshwater are as follows:

$$r_{0s54\text{–}77} = 0.7122$$

$$r_{0f54\text{–}77} = 0.6177$$

3.3.1.1.2 Correlation Analysis of the Total Aquatic Production and the Production of Seawater and Freshwater in 1978–1984 and 1985–1997

Similarly, the correlation coefficients between the total production of aquatic products and the production of seawater and freshwater in 1978–1984 and 1985–1997 can be obtained, and the results are shown in Tables 3.9 and 3.10.

The correlations between the total production of aquatic products and the production of seawater and freshwater are as follows:

$$r_{0s78\text{–}84} = 0.8163$$

Table 3.8 Correlation coefficients between the total production of aquatic products and the production of seawater and freshwater from 1954 to 1977 (Chen and Zhou 2002b)

Year	1954	1955	1956	1957	1958	1959	1960	1961
r_{0s}	1.00	0.824	0.858	0.931	0.999	0.972	0.960	0.956
r_{0f}	1.00	0.752	0.796	0.897	0.999	0.953	0.940	0.584
Year	1962	1963	1964	1965	1966	1967	1968	1969
r_{0s}	0.842	0.773	0.771	0.746	0.668	0.636	0.685	0.675
r_{0f}	0.775	0.687	0.685	0.655	0.565	0.531	0.584	0.573
Year	1970	1971	1972	1973	1974	1975	1976	1977
r_{0s}	0.631	0.576	0.505	0.531	0.476	0.471	0.458	0.436
r_{0f}	0.525	0.467	0.397	0.422	0.369	0.365	0.353	0.333

Table 3.9 Correlation coefficients between the total production of aquatic products and the production of seawater and freshwater from 1978 to 1984 (Chen and Zhou 2002b)

Year	1978	1979	1980	1981	1982	1983	1984
r_{0s}	1.0000	0.9127	0.8681	0.8142	0.7857	0.7044	0.6293
r_{0f}	1.0000	0.7548	0.6597	0.5635	0.5192	0.4124	0.3333

Table 3.10 Correlation coefficients between the total production of aquatic products and the production of seawater and freshwater between 1985 and 1997 (Chen and Zhou 2002b)

Year	1985	1986	1987	1988	1989	1990	1991
r_{0s}	1.000	0.745	0.680	0.606	0.641	0.653	0.882
r_{0f}	1.000	0.666	0.591	0.534	0.549	0.562	0.835
Year	1992	1993	1994	1995	1996	1997	
r_{0s}	0.868	0.798	0.546	0.423	0.438	0.572	
r_{0f}	0.818	0.729	0.450	0.333	0.347	0.476	

$$r_{0f78\text{–}84} = 0.6061$$

The correlations between the total production of aquatic products and the production of seawater and freshwater are as follows:

$$r_{0s85\text{–}97} = 0.6814$$

$$r_{0f85\text{–}97} = 0.6072$$

3.3.1.2 Correlation Analysis of Seawater and Freshwater Production and Fishing and Aquaculture Production

3.3.1.2.1 Correlation Analysis of Seawater Production and Fishing and Aquaculture Production

Using seawater production as the parent sequence and fishing and aquaculture production as the subsequences, the correlation coefficients between seawater production and marine fishing and mariculture production in the three time periods of 1954–1977, 1978–1984 and 1985–1997 were obtained (Tables 3.11, 3.12, and 3.13).

The correlations between seawater production and the production of marine fishing and mariculture are as follows:

$$r_{\text{sf}54\text{–}77} = 0.9769$$

$$r_{\text{sa}54\text{–}77} = 0.7830$$

Table 3.11 Correlation coefficients between seawater production and marine fishing and mariculture production from 1954 to 1977 (Chen and Zhou 2002b)

Year	1954	1955	1956	1957	1958	1959	1960	1961
r_{sf}	1.0000	0.9983	0.9711	1.0000	0.9840	0.9917	0.9979	0.9996
r_{sa}	1.0000	0.9747	0.6931	0.9996	0.8050	0.8896	0.9702	0.9935
Year	1962	1963	1964	1965	1966	1967	1968	1969
r_{sf}	0.9958	0.9866	0.9716	0.9846	0.9883	0.9989	0.9841	0.9818
r_{sa}	0.9404	0.8319	0.6972	0.8117	0.8503	0.9841	0.8066	0.7839
Year	1970	1971	1972	1973	1974	1975	1976	1977
r_{sf}	0.9733	0.9553	0.9527	0.9751	0.9710	0.9554	0.9467	0.8814
r_{sa}	0.7102	0.5897	0.5753	0.7250	0.6926	0.5906	0.5444	0.3333

Table 3.12 Correlation coefficients of seawater production and marine fishing and mariculture production from 1978 to 1984 (Chen and Zhou 2002b)

Year	1978	1979	1980	1981	1982	1983	1984
r_{sf}	1.0000	0.9673	0.9321	0.9041	0.9181	0.8458	0.7777
r_{sa}	1.0000	0.8089	0.6623	0.5740	0.6157	0.4394	0.3333

Table 3.13 Correlation coefficients of seawater production and marine fishing and mariculture production from 1985 to 1997 (Chen and Zhou 2002b)

Year	1985	1986	1987	1988	1989	1990	1991
r_{sf}	1.0000	0.9951	0.9838	0.9627	0.9579	0.9615	0.9497
r_{sa}	1.0000	0.9765	0.9253	0.8405	0.8232	0.8362	0.7943
Year	1992	1993	1994	1995	1996	1997	
r_{sf}	0.9248	0.8912	0.8845	0.8601	0.7098	0.7101	
r_{sa}	0.7154	0.6261	0.6101	0.5568	0.3333	0.3337	

The correlations between seawater production and the production of marine fishing and mariculture are as follows:

$$r_{sf78\text{–}84} = 0.9064$$

$$r_{sa78\text{–}84} = 0.6334$$

The correlations between seawater production and the production of marine fishing and mariculture are as follows:

$$r_{sf85\text{–}97} = 0.9070$$

$$r_{sa85\text{–}97} = 0.7209$$

3.3.1.2.2 Correlation Analysis of Freshwater Production and Fishing and Aquaculture Production

Using freshwater production as the parent sequence and freshwater fishing and aquaculture production as the subsequences, the correlation coefficients between freshwater production and fishing and aquaculture production in the three time periods of 1954–1977, 1978–1984, and 1985–1997 were obtained (Tables 3.14, 3.15, and 3.16).

The correlations between freshwater production and freshwater fishing and freshwater aquaculture production are as follows:

$$r_{\text{ff54–77}} = 0.6971$$

$$r_{\text{fa54–77}} = 0.5226$$

The correlations between freshwater production and freshwater fishing and freshwater aquaculture production are as follows:

Table 3.14 Correlation coefficients between freshwater production and the yields of freshwater catch and freshwater cultivation from 1954 to 1977 (Chen and Zhou 2002b)

Year	1954	1955	1956	1957	1958	1959	1960	1961
r_{ff}	1.0000	0.9021	0.9099	0.7086	0.6959	0.6929	0.7784	0.6309
r_{fa}	1.0000	0.8015	0.8157	0.5159	0.5007	0.4971	0.6061	0.4282
Year	1962	1963	1964	1965	1966	1967	1968	1969
r_{ff}	0.8706	0.8397	0.8096	0.6943	0.6702	0.6751	0.6690	0.6326
r_{fa}	0.7466	0.6966	0.6507	0.4988	0.4710	0.4765	0.4696	0.4300
Year	1970	1971	1972	1973	1974	1975	1976	1977
r_{ff}	0.6170	0.5959	0.5933	0.5867	0.5538	0.5352	0.5408	0.5282
r_{fa}	0.4138	0.3925	0.3899	0.3834	0.3523	0.3354	0.3403	0.3291

Table 3.15 Correlation coefficients between freshwater production and the yields of freshwater catch and freshwater cultivation from 1978 to 1984 (Chen and Zhou 2002b)

Year	1978	1979	1980	1981	1982	1983	1984
r_{ff}	1.0000	0.9063	0.9165	0.7917	0.5368	0.4819	0.3333
r_{fa}	1.0000	0.9613	0.9658	0.9072	0.7488	0.7052	0.5625

Table 3.16 Correlation coefficients of freshwater production and the production of freshwater catch and freshwater cultivation from 1985 to 1997 (Chen and Zhou 2002b)

Year	1985	1986	1987	1988	1989	1990	1991
r_{ff}	1.000	0.8311	0.7289	0.7016	0.7474	0.7223	0.9630
r_{fa}	1.000	0.961	0.9309	0.9217	0.9368	0.9287	0.9924
Year	1992	1993	1994	1995	1996	1997	
r_{ff}	0.6381	0.5137	0.4070	0.3654	0.4029	0.3333	
r_{fa}	0.8983	0.8410	0.7746	0.7425	0.7716	0.7146	

$$r_{ff78-84} = 0.7095$$

$$r_{fa78-84} = 0.8358$$

The correlations between freshwater production and freshwater fishing and freshwater aquaculture production are as follows:

$$r_{ff85\text{–}97} = 0.6427$$

$$r_{fa85\text{–}97} = 0.8780$$

3.3.1.3 Correlation Analysis of the Production of Seawater and Freshwater and the Production of Each Major Species

3.3.1.3.1 Correlation Analysis of Seawater Production and the Production of Each Major Species

Using the seawater yield as the mother sequence and the yield of each major species as the subsequence, the correlations between the seawater yield and the marine fish, shrimp and crabs, shellfish and algae from 1954 to 1977 are as follows:

$$r_{sf54\text{–}77} = 0.6918$$

$$r_{ssc54\text{–}77} = 0.6963$$

$$r_{sc54\text{–}77} = 0.6781$$

$$r_{sa54\text{–}77} = 0.6740$$

The correlations between seawater production and fish, shrimp and crabs, and shellfish and algae from 1978 to 1984 are as follows:

$$r_{sf78\text{–}84} = 0.8099$$

$$r_{ssc78\text{–}84} = 0.8450$$

$$r_{sc78\text{–}84} = 0.7750$$

$$r_{sa78\text{–}84} = 0.5880$$

The correlations between seawater production and fish, shrimp and crabs, and shellfish and algae from 1985 to 1997 are as follows:

$$r_{sf85\text{–}97} = 0.9345$$

$$r_{\mathrm{ssc}85\text{–}97} = 0.9604$$

$$r_{\mathrm{sc}85\text{–}97} = 0.9549$$

$$r_{\mathrm{sa}85\text{–}97} = 0.7170$$

3.3.1.3.2 Correlation Analysis of Freshwater Production and the Production of Major Species

Using freshwater production as the parent sequence and the production of each major freshwater species as the subsequence, the correlations between freshwater production and freshwater fish, shrimp and shellfish in 1954–1977 are as follows:

$$r_{\mathrm{ff}54\text{–}77} = 0.6063$$

$$r_{\mathrm{fsc}54\text{–}77} = 0.6418$$

$$r_{\mathrm{fc}54\text{–}77} = 0.7230$$

The correlations between freshwater production and freshwater fish, shrimp and crabs, and shellfish from 1978 to 1984 are as follows:

$$r_{\mathrm{ff}78\text{–}84} = 0.6661$$

$$r_{\mathrm{fsc}78\text{–}84} = 0.6620$$

$$r_{\mathrm{fc}78\text{–}84} = 0.7269$$

The correlations between freshwater production and freshwater fish, shrimp, crabs, and shellfish from 1985 to 1997 are as follows:

$$r_{\mathrm{ff}85\text{–}97} = 0.7779$$

$$r_{\mathrm{fsc}85\text{–}97} = 0.7745$$

$$r_{\mathrm{fc}85\text{–}97} = 0.8164$$

3.3.1.4 Main Conclusions

1. The different yields of the three time periods were analyzed using the gray correlation method, and the results are shown in Figs. 3.1, 3.2, and 3.3.
2. Figure 3.1 shows that the contribution of seawater production to the total production of aquatic products is always greater than that of freshwater production. However, with the passage of time, the contribution of seawater production to the

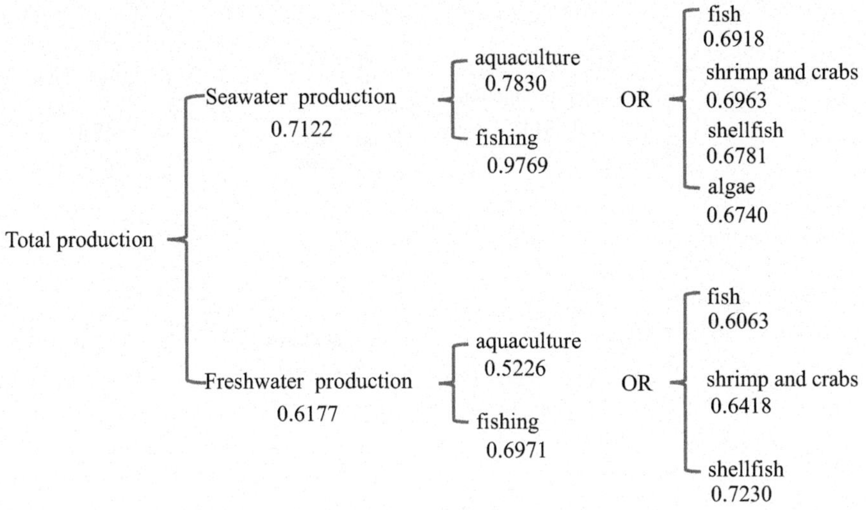

Fig. 3.1 Correlation between fishery production in China from 1954 to 1977 (Chen and Zhou 2002b)

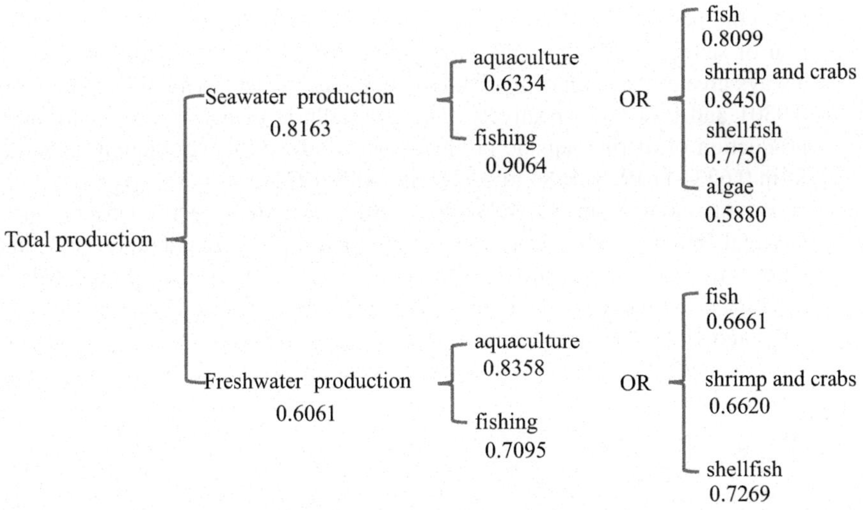

Fig. 3.2 Correlation between fishery production in China from 1978 to 1984 (Chen and Zhou 2002b)

total production of aquatic products decreased, and its correlation degree decreased from 0.8163 in 1978–1984 to 0.6814 in 1985–1977, while the contribution of freshwater production to the total production of aquatic products decreased. The correlation degree increased slightly from 0.6061 in 1978–1984 to 0.6072 in 1985–1977.

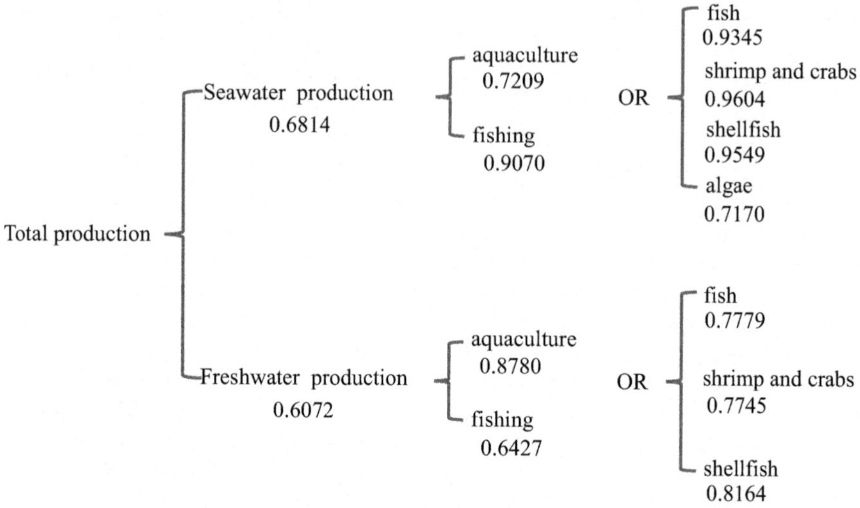

Fig. 3.3 Correlation between fishery production in China from 1985 to 1997 (Chen and Zhou 2002b)

3. From the perspective of the composition of seawater production, the contribution of seawater fishing to the total production of marine aquatic products is always greater than that of mariculture, and the average correlation degrees are 0.9301 and 0.7124, respectively. The contribution of mariculture to the total production of marine aquatic products increased, and its correlation increased from 0.6334 in 1978–1984 to 0.7209 in 1985–1977.

   For marine fish, shrimp and crabs, shellfish and algae, their contributions to the total aquatic product production were basically the same before 1977, and their correlations were all between 0.674 and 0.696, indicating that fishery resource development and utilization are relatively balanced. However, between 1978 and 1984, the contribution of fish, shrimp, and crabs to the production of seawater aquatic products increased significantly, and their correlations increased to 0.8099 and 0.8450, respectively, while the correlations between algae and the production of seawater aquatic products decreased to 0.5880. After 1984, the increase in the production of seawater aquatic products mainly came from shrimp, crabs, shellfish and fish, and the correlations were 0.9604, 0.9549, and 0.9345, respectively.
4. In freshwater fisheries, fishing and aquaculture have undergone structural changes. Before 1977, freshwater production mainly came from freshwater fishing, with a correlation degree of 0.6971, while the correlation degree between freshwater aquaculture and freshwater aquatic product production was only 0.5226. From 1978 to 1984, both freshwater fishing and freshwater aquaculture developed to a certain extent. However, the development of freshwater aquaculture was even more rapid, and its correlation reached 0.8358. After 1985, the development of freshwater fishing shrank, its contribution to the

production of freshwater aquatic products decreased, and freshwater aquaculture was further developed. and its correlation reached 0.8780.

Among freshwater fish, shrimp, and shellfish, the contribution of shellfish to the production of freshwater aquatic products is always highest, while the contribution of fish, shrimp, and crabs is relatively low. In different time periods, the contributions of freshwater fish, shrimp, crabs, and shellfish to their yields increased, and the correlations between 1985 and 1997 were 0.7779, 0.7745, and 0.8164, respectively.

5. The correlation analysis of the time series shows that it is impossible to increase the proportion of marine catch production in marine fisheries. In contrast, the potential of marine aquaculture is relatively large, but attention should be given to the protection of the aquaculture water environment. In addition, basic research on the aquaculture capacity and optimal combination of waters has been carried out. In addition, in freshwater fisheries, the increase in freshwater fishing yield is also difficult to achieve. Some resources have been overfished, and freshwater aquaculture has been the main source of the recent increase in the production of aquatic products.

   Through the above analysis, we can clearly understand the structure of China's fishery production in the past and present as well as its status, existing problems and potential, which further indicates that the development of China's fisheries must be based on natural resource conditions and the level of fishery economic development. We should formulate development plans in a targeted manner, strive to reduce fishing intensity, and focus on the development of some ecological fishery economies to ensure the sustainable development of China's fisheries.

3.3.1.5 Application of Gray Correlation in the Field of Fishery Resource Assessment

3.3.1.5.1 Selection of Factors Affecting Changes in Fishery Resources

Yan et al. (1996) published "Relational factors for changes in Taihu silverfish resources and methods for resource forecasting." The gray correlation analysis method was used to screen the main factors related to the Taihu silverfish resources. According to the qualitative analysis of production practices, the nonbiological factors that affect the changes in the number of whitebait in Lake Taihu are mainly the water level in spring from March to May, the water level in summer from June to September, and the fishing intensity (based on the amount of fishing in the lake after spring and autumn floods). In terms of biology, there is a relationship between competing bait, predation, and being preyed upon. Therefore, in the quantitative analysis, the yields of whitebait, lake anchovy, shrimp, culter, and small trash fish are considered the main sequences. The six items, including the amount of whitebait in spring and autumn, were used as subsequences to calculate the degree of gray correlation. The original values are shown in Table 3.17.

Table 3.17 The yield of natural fish in Lake Taihu and the values of its main factors (Yan et al. 1996)

Content		1989	1990	1991	1992	1993	1994
Main sequences	Whitebait (tons)	1509.06	1479.27	2008.62	1606.92	1763.50	1118.18
	Lake anchovy (ton)	7460.48	8142.34	6634.16	4625.41	3486.50	6706.55
	Shrimp (tons)	898.22	1122.59	751.35	560.50	523.10	879.40
	Culter (ton)	518.30	551.35	731.62	732.66	922.25	301.95
	Small trash fish (tons)	2039.29	1884.88	2.070.88	3102.21	4057.87	2827.57
Subsequences	Ship tonnage (ton)	31173.5	31462.5	31987.5	43721.0	45695.5	38276.0
	Labor (person)	11083.0	11382.0	11429.0	11843.0	12546.0	12386.0
	The resource index of whitebait in spring	3.48	2.70	5.05	3.45	2.55	4.3100
	The resource index of whitebait in autumn	0.13	0.50	0.90	0.80	0.13	0.0600
	Water level in spring (m)	3.02	3.14	3.33	3.13	3.08	2.9800
	Water level in autumn (m)	3.52	3.17	3.96	3.03	3.70	2.9325

Note: The unit of the resource index of whitebait is kg/h

Table 3.18 Correlation values of natural fish and their related factors and their order (Yan et al. 1996)

Fish	Ship tonnage	Labor	The resource index of whitebait in spring	The resource index of whitebait in autumn	Water level in spring	Water level in autumn
Whitebait	0.7528(4)	0.8553(2)	0.7484(5)	0.4565(6)	0.8501(3)	0.8978(1)
Lake anchovy	0.5741(5)	0.6745(3)	0.6678(4)	0.4896(6)	0.7218(2)	0.7349(1)
Shrimp	0.5600(5)	0.6665(1)	0.6332(4)	0.4132(6)	0.6584(2)	0.6541(3)
Culter	0.8328(1)	0.7927(2)	0.7208(5)	0.4945(6)	0.7900(3)	0.7565(4)
Small trash fish	0.8664(1)	0.7375(2)	0.6785(5)	0.5146(6)	0.7050(3)	0.6866(4)

Note: The numbers in parentheses are the order of the degree of gray correlation

The data in Table 3.17 were used to calculate the general gray correlation degree, and the resolution coefficient $\rho = 0.55$ was used to obtain the gray correlation degree value in Table 3.18. Table 3.18 shows that the water level is the first correlation factor for whitebait and lake anchovy, and fishing intensity is the first correlation factor for shrimp, culter, and small trash fish.

3.3.1.5.2 Selection of Factors Affecting the Catch

Wang (1996) published “Gray correlation analysis of Danjiangkou Reservoir catch and its related factors.” The raw data of release specification for fish, fishery law enforcement management, the fish release quantity, the amount of fishing effort, the inflow of water, and the reservoir area are collected (Table 3.19). The study was divided into two time periods: 1971–1982 (case I) and 1982–1990 (case II).

Assume that the data series of catch, release specification for fish, fishery law enforcement management, fish release quantity, fishing effort, inflow of water, and reservoir area after the averaged values of each year are x_0, x_1, x_2, x_3, x_4, x_5, and x_6, respectively, the correlation coefficient corresponding to each period is $\xi_i(k)$, and the correlation degree is r_i $(i = 1, 2..., 6)$. The results of the averaging of the raw data are shown in Table 3.20. The correlation coefficient is shown in Table 3.21. Table 3.22 shows the degree of gray correlation and its ranking.

Table 3.22 shows that in case I, the correlation sequence between each factor and the catch is 0.992 > 0.988 > 0.984 = 0.984 > 0.871 > 0.752, i.e., $r_1 > r_6 > r_4 = r_5 > r_2 > r_3$, r_1, r_6, r_4, r_5, r_2, and r_3 are the gray correlation between the release specification for fish (cm) and the catch, reservoir area (10,000 hm^2) and the catch, fishing effort (10,000 tons) and the catch, inflow of water (100 million m^3) and the catch, fishery law enforcement management (10,000 yuan) and the catch, and fish release quantity and the catch, respectively.

The gray correlation study showed the following:

Table 3.19 Raw data of gray correlation analysis (Wang 1996)

Year	Release specification for fish (cm)	Fishery law enforcement management (ten thousand yuan)	Fishing effort (10,000 tons)	Fish releasing quantity (ten thousand kg)	Inflow of water (100 million m^3)	Reservoir area (10,000 hm^2)	Catch (10,000 kg)
1971	8	2	14	0.135	377.47	5.10	90.42
1972	8	2	13.2	0.135	281.25	5.09	97.93
1978	12.8	15.83	22.5	2.175	228.28	3.74	96.41
1979	12.8	15.83	22.5	2.175	320.41	4.07	107.85
1986	15	39.75	58	10.04	240.11	5.11	227.23
1987	16	37.20	70.02	10.04	375.39	5.40	227.40
1988	20	26.92	68	7.07	380.75	5.00	263.53
1989	20	17.38	59.9	7.07	375.00	4.80	237.00
1990	20.4	14.17	72	3.09	389.00	5.80	278.00

Table 3.20 Initialization results of raw data (initialization using 1971 as the denominator) (Wang 1996)

Situation	Year	x_0	x_1	x_2	x_3	x_4	x_5	x_6
I	1971	1.0	1	1	1	1	1	1
	1972	1.0	1	1	1	0.94	0.75	1
	1978	1.07	1.60	7.92	16.11	1.61	0.61	0.73
	1979	1.19	1.60	7.92	16.11	1.86	0.85	0.80
	1986	2.51	2	19.88	74.37	4.14	0.64	1
II	1986	1	1	1	1	1	1	1
	1987	1	1	0.94	1	1.21	1.56	1.06
	1988	1.15	1.25	0.68	0.70	1.03	1.59	0.98
	1989	1.04	1.25	0.45	0.70	1.03	1.56	0.94
	1990	1.22	1.28	0.36	0.31	1.24	1.62	1.14

Table 3.21 Correlation coefficients between various factors and catches (Wang 1996)

Situation	Year	ξ_1	ξ_2	ξ_3	ξ_4	ξ_5	ξ_6
I	1971	1	1	1	1	1	1
	1972	0.998	0.998	0.998	0.996	0.991	0.998
	1978	0.985	0.840	0.705	0.985	0.987	0.991
	1979	0.989	0.842	0.721	0.982	0.991	0.989
	1986	0.986	0.674	0.336	0.957	0.951	0.960
II	1986	1	1	1	1	1	1
	1987	1	0.883	1	0.684	0.448	0.883
	1988	0.820	0.492	0.503	0.791	0.508	0.728
	1989	0.684	0.435	0.572	0.978	0.467	0.820
	1990	0.883	0.346	0.333	0.958	0.532	0.850

Table 3.22 The degree of gray correlation and ranking of each factor and the catch (Wang 1996)

Situation		r_1	r_2	r_3	r_4	r_5	r_6
Gray relational degree	I	0.992	0.871	0.752	0.984	0.984	0.988
	II	0.877	0.631	0.682	0.882	0.591	0.856
Sorting	I	1	5	6	3	3	2
	II	2	5	4	1	6	3

1. The release specification for fish had the greatest impact on the catch. The Danjiangkou Reservoir is a type of reservoir mainly inhabited by Erythroculter. When the release specification for fish is small, most of the fish species will be eaten by Erythroculter.
2. The next impact factor on the catch is the reservoir area. In a reservoir with a small area, fish will be restricted by the density factor, so the population density may be very high, but the total yield is not high. The larger the area of the reservoir is, the larger the living space of the fish. The density decreases, the food

can increase relatively, the fish grows rapidly, and the individual is large; therefore, the total yield is high.

3. Fishing effort and inflow of water also have a greater impact on the catch, which is second only to the reservoir area and has a greater impact on the catch than fishery law enforcement management and the fish release quantity. The inflow of water indicates the amount of nutrients, which directly or indirectly restricts the amount of fish resources and growth.
4. The impact of fishery law enforcement management and the fish release quantity on the catch is small. The Danjiangkou Reservoir has a relatively large area, so it is difficult to achieve effective fishery law enforcement management. This is the reason why r_2 is smaller than the other factors. However, $r_2 = 0.887 > 0.8$, indicating that fishery law enforcement management is still closely related to catch. Therefore, it is necessary to strengthen fishery law enforcement management. In case I, r_3 was the smallest, indicating that the release specification for fish (<16 cm) was too small, and only a one-sided increase in the fish release quantity without increasing the release specification for fish resulted in a very poor economic benefit.

In case II, there are $0.882 > 0.877 > 0.856 > 0.682 > 0.631 > 0.591$, i.e., $r_4 > r_1 > r_6 > r_3 > r_2 > r_5$. The gray correlation analysis shows the following:

1. Under this situation, fishing effort has replaced the release specification for fish as the primary factor affecting the catch, indicating that enhanced fishing is economical. The release specification for fish dropped to second place, indicating that the release specification for fish (20.4 cm) at this time basically met the requirements. If the release specification for fish continues to increase, the effect on the catch will gradually weaken, and thus, the economic effect will be worse.
2. The effect of the fish release quantity on the catch increased from the sixth to fourth, indicating that the effect of the fish release quantity under this size gradually increased, and an increase in the fish release quantity significantly increased the catch. It is economical to increase the fish release quantity.
3. The inflow of water decreased from the fourth to the last, indicating that the effect of fish release quantity on the catch in this period replaced the position of the inflow of water; that is, the fish release quantity (10,000 kg) can affect the catch compared to the inflow of water (100 million m^3).
4. Fishery law enforcement management is still in fifth place, indicating that the role of the fishery in the two situations has not changed. If the fish release quantity cannot be increased, the fishing effort is certain, and fishery law enforcement management should be strengthened to obtain a higher catch.
5. It is understandable that the reservoir area is always in a more important position. Therefore, illegal occupation of the water surface should be minimized, or the water surface occupation fee should be appropriately levied.

3.3.1.5.3 Evaluation of the Abundance of Fishery Resources

Ommastrephes bartramii, neon flying squid, is widely distributed in the North Pacific Ocean and has abundant resources. This resource was first developed and utilized by Japanese squid fishing survey vessels in 1974. Subsequently, South Korea and Taiwan Province of China also joined the development ranks and gradually developed into a fleet of gill net-based fishing fleets. The output of squid reached 300,000–400,000 tons. In 1993, the Chinese mainland began to develop and utilize the resources of neon flying squid in the North Pacific Ocean, and the operating fishing grounds continued to expand eastward. The squid fishing fisheries in the North Pacific Ocean have become an important part of China's offshore fisheries. Therefore, conducting research on the spatial distribution of its resources and fishing grounds and their interannual change is of great significance to ensure the rational exploitation and utilization of this squid (Chen et al. 2003).

Chen et al. (2003) used the statistics of China's squid fishing catch in the North Pacific Ocean from 1995 to 2001, including the operating location, operating date, number of vessels operating in the fishing area, total production in the fishing area, and average daily production. The average daily production for each longitude and latitude in each year was also calculated. The gray relational degree is used to evaluate the resource status of neon flying squid in each year. That is, the maximum value of each longitude (latitude) in each year is selected to form the optimal vector and used as the parent sequence. The greater the degree of gray correlation is, the better the resource status, and vice versa.

Based on the history of Chinese squid fishing vessels in the North Pacific Ocean exploiting and utilizing squid resources, this study was conducted in three sea areas to the west of 160°E, 160°E to 170°E and to the east of 170°E. The results of this study are as follows:

1. Comparison of the resource status of squid in the waters west of 160°E

   In the sea area west of 160°E, the optimal vectors for the average daily yield of each longitude (143°E–160°E) are 2.2, 2.848, 3.182, 3.719, 5.803, 2.524, 3.526, 2.406, 2.462, 2.317, 2.667, and 3.806. The gray correlations between the average daily production of each year at different longitudes and the optimal average daily production in 1995–2001 were 0.7061, 0.5555, 0.6851, 0.8025, 0.7097, 0.6294, and 0.5504, respectively. The order of abundance of squid from high to low in each year was 1998, 1999, 1995, 1997, 2000, 1996, and 2001.

2. Comparison of the resource status of squid in the waters west of 160°E–170°E

   In the sea area of 160°E–170°E, the optimal vectors of the average daily yield of each longitude (161°E–170°E) from 1997 to 2001 were 1.779, 2.013, 2.371, 2.048, 1.873, 1.569, 1.585, 1.453, and 1.549, respectively. The gray correlations between the average daily production of each year at different longitudes and the optimal average daily production in 1997–2001 were 0.6620, 0.8467, 0.6318, 0.6040, and 0.4938, respectively. The order of abundance of squid from high to low in each year is 1998, 1997, 1999, 2000, and 2001.

3. Comparison of the resource situation of squid in the sea east of 170°E

In the sea area to the east of 170°E, the optimal vectors of the average daily yield of each longitude (171°E–173°W) from 1999 to 2001 are 1.178, 0.857, 1.346, 1.706, 1.817, 1.876, 1.224, 1.454, 1.07, and 1.427, 1.375, 1.421, 1.587, 1.642, 1.627, 1.276, and 0.819. The gray correlations between the average daily production of each year at different longitudes and the optimal average daily production in 1999–2001 were 0.966, 0.640, and 0.677, respectively. The order of abundance of squid from high to low in each year is 1999, 2001, and 2000.

4. Comparison of the resource status of squid at different latitudes

   In the North Pacific Ocean, the optimal vectors at various latitudes (37°N–45° N) are 2.05, 4.152, 2.496, 2.225, 2.895, 2.752, 2.405, 2.693, and 2.927. The gray correlations between the yield of each year and the optimal average daily yield were 0.854, 0.709, 0.768, 0.861, 0.769, 0.649, and 0.630, respectively. The order of abundance of squid from high to low in each year is 1998, 1995, 1999, 1997, 1996, 2000, and 2001.

Based on the gray correlation evaluation, we obtained the resource status of squid in various sea areas of the North Pacific Ocean from 1995 to 2001. The status of squid resources in the North Pacific Ocean was the best in 1998, while the status of squid in 2000, 2001, and 1996 was poor. In 1996, it was at an intermediate level. This is basically consistent with the actual production situation and marine environmental conditions. For example, in 1998, the Kuroshio power was strong, and it was a warm-water year, while in 1996, the Oyashio was strong and the Kuroshio was relatively weak, and it was a cold-water year. Therefore, as a warm-water species of this squid, the strength of the Kuroshio directly affects the amount of resources and the formation of fishing grounds for neon flying squid.

3.3.2 Application of Gray Correlation in the Evaluation of Sustainable Use of Fishery Resources

The sustainable use of fishery resources is the core and essential issue of the sustainable development of the fishery economy. Chen and Zhou (2002a) published the "Gray Relational Assessment of the Sustainable Utilization of Fishery Resources" and analyzed the evaluation of the sustainable use of fishery resources in the East China Sea using the gray relevance analysis method.

The evaluation index system proposed by Chen and Zhou (2002a) includes the three subsystems of resource environment, society, and economy. The resource environment subsystem includes trophic level R_{101}, the proportion of high-quality fish in marine fishing yield R_{102}, the proportion of catch from nonselective fishing gear in the total marine fishing yield R_{103}, the marine fishing yield per ship R_{104}, the marine fishing yield per tonnage of motor-driven fishing vessels R_{105}, and the marine fishing yield per kilowatt of motor-driven fishing vessels R_{106}. In the subsystems of society, there are six indicators, including marine fishing professional labor R_{201}, marine fishing part-time labor R_{202}, the proportion of marine fishing labor in fishery

Table 3.23 Index values of the resource and environmental subsystems after screening (Chen and Zhou 2002a)

Year	R_{101}	R_{102} (%)	R_{103}(%)	R_{106} (*t*/KW)
1978	2.64	63.189	0.436	1.178
1979	2.72	59.118	0.411	1.050
1980	2.73	46.483	0.569	1.038
1981	2.72	51.056	0.585	0.956
1982	2.64	48.178	0.622	0.935
1983	2.63	38.596	0.645	0.875
1984	2.54	41.034	0.677	0.891
1985	2.56	39.083	0.623	0.869
1986	2.52	37.618	0.676	0.881
1987	2.50	37.917	0.671	0.821
1988	2.40	30.400	0.683	0.727
1989	2.43	36.130	0.699	0.683
1990	2.49	36.125	0.724	0.663

labor R_{203}, the proportion of marine fishing labor in fishery population R_{204}, fishery population R_{205}, and the per capita share of aquatic products R_{206}. The subsystems of the economy include 11 indicators, including marine fishing yield R_{301}, the proportion of marine fishing yield in marine fishery yield R_{302}, the proportion of marine fishing yield in the total fishery output R_{303}, the proportion of the total fishery output in the agricultural output value R_{304}, the number of motor fishing vessels R_{305}, the total tonnage of motor-driven fishing vessels R_{306}, the total power of motor-driven fishing vessels R_{307}, the per capita income of fishermen R_{308}, the per capita income of fishermen R_{309}, the per capita marine fishing yield of fishing labor R_{310}, and the per capita marine fishing yield of fishery population R_{311}. The data are from the "China Fishery Statistics Collection" (1989–1993) (edited by the Fishery Bureau of the Ministry of Agriculture of the People's Republic of China, 1996), the China Fishery Statistics Collection (1994–1998) (edited by the Fishery Bureau of the Ministry of Agriculture of the People's Republic of China, 2000), and China Fisheries Statistics for 40 years (edited by the Fisheries Department of the Ministry of Agriculture of the People's Republic of China, 1991).

Considering that there are differences in the dimensions of the original data and the significant differences in the order of magnitude between the indicators, the correlation coefficients between the indicators in each subsystem should be initialized. Additionally, 11 indicators for evaluation were obtained through principal component analysis and independence analysis. They are R_{101}, R_{102}, R_{103}, and R_{106}; R_{201}, R_{203}, R_{204}, and R_{206}; and R_{302}, R_{310}, and R_{311}. The raw data of each indicator and their weights are shown in Tables 3.23, 3.24, 3.25, and 3.26.

The optimal value of each indicator was selected to form the parent sequence, and each year was used as the subsequence. The general gray correlation analysis method was used to evaluate the sustainable use of fishery resources in the East China Sea between 1978 and 1990. The correlation coefficient was set to 0.5. The results are shown in Table 3.27.

Table 3.24 The values of each index of the social subsystem after screening (Chen and Zhou 2002a)

Year	R_{201} (people)	R_{203} (%)	R_{204} (%)	R_{206} (people)
1978	341,835	75.146	21.636	1,830,102
1979	344,977	68.925	21.266	1,887,286
1980	359,779	61.343	22.191	1,940,048
1981	381,907	61.467	22.757	1,992,514
1982	366,601	52.893	19.992	2,106,660
1983	320,813	44.588	18.063	2,175,966
1984	344,289	50.804	21.316	1,966,558
1985	361,232	50.216	20.325	2,302,175
1986	408,852	52.307	21.830	2,333,331
1987	427,508	52.343	22.719	2,352,121
1988	442,793	51.419	23.305	2,423,091
1989	416,503	50.048	21.748	2,417,011
1990	419,822	48.973	20.757	2,524,825

Table 3.25 Indicator values of the economic subsystem after screening (Chen and Zhou 2002a)

Year	R_{302} (%)	R_{310} (*t*/fishing labor)	R_{311} (*t*/people)
1978	91.433	3.839	0.831
1979	89.969	3.540	0.753
1980	89.582	3.345	0.744
1981	89.386	3.332	0.730
1982	88.706	3.485	0.726
1983	87.590	3.731	0.678
1984	85.479	3.550	0.814
1985	84.333	3.646	0.731
1986	84.643	3.526	0.779
1987	84.289	3.674	0.858
1988	83.179	3.674	0.837
1989	83.589	4.148	0.894
1990	84.068	5.472	0.910

Table 3.26 The weight of each subsystem and its index (Chen and Zhou 2002a)

Indicator	R_1	R_2	R_3	
Weight	0.5245	0.1688	0.3067	
Indicator	R_{101}	R_{102}	R_{103}	R_{106}
Weight	0.1621	0.1574	0.2537	0.4267
Indicator	R_{201}	R_{203}	R_{204}	R_{206}
Weight	0.2857	0.2857	0.2857	0.1429
Indicator	R_{302}	R_{310}	R_{311}	
Weight	0.3333	0.3333	0.3333	

From the evaluation results in Tables 3.27, it can be seen that the level of sustainable utilization of fishery resources in the East China Sea between 1978 and 1990 basically showed a downward trend. Among them, the level of sustainable use during 1983 to 1986 was relatively low, the level of sustainable use in 1983 was the

Table 3.27 Evaluation results of the sustainable use of fishery resources in the East China Sea from 1978 to 1990 (Chen and Zhou 2002a)

Year	1978	1979	1980	1981	1982	1983	1984
Evaluation value	0.7473	0.6541	0.5579	0.5401	0.4703	0.4234	0.4347
Year	1985	1986	1987	1988	1989	1990	
Evaluation value	0.4330	0.4550	0.4915	0.5165	0.4691	0.5252	

lowest, and the level of sustainable use in 1990 was only 75% of that in 1978. This evaluation result is basically in line with the reality of the exploitation and utilization of fishery resources in the East China Sea. Since the 1980s, the fishery resources in the East China Sea have been overexploited and utilized. Especially in 1983, due to the decline in offshore fishery resources, some resources in the open sea, such as mackerel and mackerel, were not exploited and utilized. After 1983, the development and utilization of some pelagic fish resources in offshore waters promoted the improvement of the level of sustainable utilization of fishery resources (Chen and Zhou 2002a).

3.3.3 Evaluation of the Influencing Factors of Fishery Water Quality

Naked carp in Qinghai Lake is the only aquatic economic animal in Qinghai Lake and plays a central role in the entire Qinghai Lake ecosystem. Therefore, it is particularly important to evaluate the aquatic environment quality of the aquatic germplasm resources of the Qinghai Lake Naked Carp. Wang (2015) used the gray relational analysis method to evaluate and analyze the fishery environmental quality based on the water quality data of five monitoring sections in the Qinghai Lake Naked Carp National Aquatic Germplasm Resource Reserve in 2012, with a view to provide a scientific basis for protecting and recovering Naked Carp resources in Qinghai Lake.

A total of five monitoring sections were set up in the study area: Shaliu River, Quanji River, Buha River, Heima River, and Wharf 151. The evaluation criteria for the relevant factors of water environmental quality refer to the "Environmental Quality Standards for Surface Water" (GB3838-2002), and the standards are shown in Table 3.28. The research data are from the Fishery Environmental Monitoring Station of Qinghai Province.

First, the study series was normalized: for the comparison series, the standard value corresponding to Class I was set to 1, and the standard value corresponding to Class V water quality was set to 0; the normalized method of taking the standard value of dissolved oxygen is contrary to other water quality parameters. The normalized comparison sequence is shown in Table 3.29. For the series to be compared, among the 5 values of the same evaluation factor, the highest pollutant content is taken as 0, the lowest is taken as 1, and the remaining 3 standard values are

Table 3.28 Water environmental quality assessment standards (Wang 2015)

Class	DO	TN	Potassium permanganate index	Ammonia	Mercury	Hexavalent chromium
I	7.5	0.2	2	0.15	0.00005	0.01
II	6	0.5	4	0.5	0.00005	0.05
III	5	1.0	6	1.0	0.0001	0.05
IV	3	1.5	10	1.5	0.001	0.05
V	2	2.0	15	2.0	0.001	0.10

Table 3.29 Comparison series after normalization (Wang 2015)

Class	DO	TN	Potassium permanganate index	Ammonia	Mercury	Hexavalent chromium
I	1	1	1	1	1	1
II	0.727	0.833	0.846	0.811	1.000	0.556
III	0.545	0.556	0.692	0.541	0.947	0.556
IV	0.182	0.278	0.385	0.270	0.000	0.556
V	0	0	0	0	0	0

Table 3.30 Normalized sequence of five sections to be compared (Wang 2015)

Monitoring point	DO	TN	Potassium permanganate index	Ammonia	Mercury	Hexavalent chromium
Liusha River	0.533	1	0.706	0.726	1	0
Quanji River	0.267	0.167	1	0.732	0.500	1
Heimahe River	0	0.208	0.941	0.701	0.750	0.056
Buha River	0.667	0.333	0.824	1	0	0.778
Wharf 151	1	0	0	0	0	0.722

Table 3.31 Calculation results of the correlation degree of each monitoring point (Wang 2015)

Class	Liusha river	Quanji river	Heimahe river	Buha river	151 Wharf
I	0.580	0.526	0.362	0.506	0.261
II	0.712	0.479	0.492	0.583	0.261
III	0.701	0.497	0.518	0.537	0.314
IV	0.302	0.495	0.453	0.571	0.567
V	0.273	0.297	0.483	0.320	0.694

determined according to the interpolation method. Table 3.30 shows the series of normalized data to be compared.

The correlation coefficient was set to 0.5, and the gray correlation degree was calculated. The calculation results are shown in Table 3.31. The higher the correlation degree is, the better the correlation with the comparison series, indicating that the water quality level is close to a certain water quality standard in the comparison series. It can be seen from the calculation results in Table 3.31 that the Shaliu River monitoring point has the highest correlation with the Class II water quality standard, but the correlation with the Class III water quality standard is also relatively large, and its correlation degree is 0.701, indicating that the water quality has just reached Class II. In the same way, the Heima River, Buha River, and 151 Wharf can be evaluated as Class III, Class II, and Class V. Among them, the correlation between the monitoring points of the Heima River and the Class II water quality standards is only second to that of the Class III water quality standards, indicating that the water environment quality presents a trend of a virtuous cycle. The correlation degree indicates that the water environment is in an unstable state, and there is a trend of deterioration to Class III water. The water environment quality of the 151 Wharf monitoring point is assessed as Class V because this monitoring point has four monitoring indicators inferior to it. For the same environmental factors at other monitoring points, when the series of comparisons were normalized, four factors in the series after normalization were 0, resulting in the evaluation of the water environment as overprotected. Taking all the monitoring points as the five influencing factors of the Qinghai Lake naked carp aquatic germplasm resource protection zone, the arithmetic mean is calculated, and the correlation between the entire water area and the water quality standards of Classes I, II, III, IV, and V is found. They were 0.477, 0.505, 0.513, 0.478, and 0.413, respectively. The overall water quality evaluation of the Qinghai Lake naked carp aquatic germplasm resource protection area is Class III, and the correlation between this area and the Class II water quality standard is only second to the Class III water quality standard, indicating the overall water quality status of the entire region.

3.3.4 Application of Gray Correlation in the Aquaculture Industry

Chen (1991) published "The application of gray correlation analysis in aquaculture." In the study, the feed coefficient column was used as the reference series, and the animal and plant protein content series were used as the comparison series (Table 3.32). The gray correlation analysis showed that the correlation degree between the plant protein content and the feed coefficient was 0.96, and the

Table 3.32 Relationship between plant and animal protein content and diet coefficient (Chen 1991)

Item	Feed group number						
	1	2	3	4	5	6	7
Plant protein content (%)	24.45	22.55	21.61	20.67	19.70	18.76	17.82
Animal protein content (%)	–	2.02	3.06	4.04	5.08	6.12	7.10
Feed coefficient	1.86	1.72	1.68	1.50	1.67	1.85	1.95

Table 3.33 Benefits of fish farming in rice fields (Chen 1991)

	Survey site number						
Item	1	2	3	4	5	6	7
Amount of fish species (tails/mu)	444	269	69	419	263	395	500
Organic fertilizer input (kg/mu)	926	96	28.5	416.5	333.5	329	416.5
Labor input (units/mu)	3.7	4.6	2.9	4.2	10.0	1.3	2.1
Net income from fisheries (yuan/mu)	83.96	39.87	30.19	54.08	86.18	40.00	61.39

correlation degree between the animal protein content and the feed coefficient was 0.58. Therefore, the effect of plant and animal protein content in the grass carp diet on the diet coefficient of grass carp was significantly stronger than that of the latter. Studies have shown that the rational selection of plant protein content is a crucial factor (Chen 1991).

At the same time, Chen (1991) used fishery net income as the reference series and fish species input, organic fertilizer input, and labor input as the comparative series (Table 3.33). The gray correlation between labor input and fishery net income was 0.76, 0.62, and 0.77, respectively. The analysis showed that the main factor affecting the net income of rice-fish farming was labor input, followed by fish species input. The key to achieving good returns from rice-fish farming is the implementation of human management measures, as described by many aquaculture experts as "three-part farming, seven-part management."

3.3.5 Applications in Fishery Biology

3.3.5.1 Analysis of Factors Affecting Fish Behavior

Gray correlation analysis is also used in the evaluation of factors affecting fish behavior. He (1989) published "A Study on the Application of Gray System Theory in the Analysis of Fish Cage Experiments," which used the gray relational analysis method to analyze the factors affecting the catch of the cage. In the cage fishery, there are many factors that affect the yield of cage catches, and the degree of influence is also different. However, the primary and secondary relationships of the factors are not clear, and the entire system can be regarded as gray. Four factors, i.e., cage time, cage space, bait in the cage, and cage darkness, were considered in the experiment. In the factor analysis, the cage catch yield was used as the reference sequence, and the four factors were used as the subsequences. Because it is difficult to measure the freshness of the bait and the darkness in the cage, fuzzy quantification was used for processing. The shading is represented by x, and the freshness is represented by y. Then, the values of the two are set as follows:

Table 3.34 Experimental data of various factors and cage catch production after normalization (He 1989)

Time series	1	2	3	4
Cage yield X_0	1.125	1.333	2.000	2.200
Fishing time X_1	0.1987	0.1987	1.391	0.9934
(Fishing time X_1)[a]	(48)[a]	(48)[a]	(24)[a]	(60)[a]
Cage space X_2	1.074	0.7395	0.6219	0.9561
Cage darkness X_3	0.7500	0.8330	1.800	1.667
Freshness of the bait X_4	0.8750	1.130	2.333	2.000

[a]The data in parentheses represent unstandardized experimental data, and the unit is hour

Table 3.35 Degree of correlation between various factors and cage catch yield (He 1989)

	Influencing factors			
Contents	Bait freshness	Darkness in the cage	Cage space	Fishing time
Gray correlation	0.8645	0.7568	0.6345	0.4997
Ranking	1	2	3	4

$$x = [2, 1, 0, 0]^T$$

Round truncated cage, folding cage, rectangular cage, round port cage

$$y = [3, 1, 1, 0, 0]^T$$

Hairtail, mackerel, anchovy, loach, eel

The obtained test data are shown in Table 3.34 after processing according to the above methods.

After calculation, the gray correlation between each factor and the catch was obtained (Table 3.35). Table 3.35 shows that the main factor affecting the catch is the freshness of the bait, indicating that the fish in the island reef area prefer to eat bait with a strong odor. Fresh bait was placed in the fish cage, and the cage feeding rate was high, which was confirmed by the production practice. At the same time, the impact of cage darkness on the yield of cage catches was also very significant, second only to the bait. This may be caused by the habit of fish in the island reef area. These fishes live in rocky reefs, caves, or crevices for a long time and are used to dark areas. It can also be seen from Table 3.35 that the effect of cage space cannot be ignored. A large cage space is conducive to fish habitation, and the probability of fish escaping from the cage is relatively small. The large cage space is also conducive to fish survival. However, oversized cages can lead to problems such as increased costs and increased operational and transportation difficulties. For the fishing time factor, the calculation results showed that it had the smallest impact on the catch of cages. This may be because the samples in the experimental data were all collected over 24 h, and the fish were likely to enter the cage within 24 h, thus causing the fishing impact of time on the catch of the cages. If the fish have not entered the cage within

24 h, the freshness of the bait in the cage will be greatly reduced, and the bait factor will not play a role, which directly affects the catching effect of the cage. Despite the extension of the cage time, the effect was not increased.

3.3.5.2 Selection and Comparison of Fish Growth Models

The gray correlation analysis method was used to select and evaluate the fish growth model and achieved good results. Chen (1991) published "The Application of Gray Relational Analysis in Fisheries," which used the gray relational analysis method to compare three commonly used fish growth models with the actual values and select the optimal model based on the degree of gray relational analysis.

According to the known growth data of grass carp, the following three growth models were obtained:

Growth model I: $W_t = 12011[1-e^{-0.3002\ (t\ +\ 0.3399)}]^{3}$
Growth model II: $W_t = 36971–38031\ e^{-0.0404t}$
Growth model III: $W_t = 7298/(1+ e^{3.9890–1.3086t})$

According to the three growth models, the theoretical body weight of each age group was calculated (Table 3.36). The measured body weight was used as the reference column, and the model-calculated value was the subsequence. The correlation between the calculated value and the measured value was obtained using the gray correlation analysis method. The gray correlations with models I, II, and III were 0.85, 0.89, and 0.77, respectively, indicating that model II was the better model for describing the body weight growth pattern of grass carp (Table 3.36).

3.3.5.3 Gray Correlation Analysis of Fish Morphological Traits and Body Weight

In the breeding of aquatic animals, the body weight of growth traits is often an important indicator for the selection of growth traits. In addition to body weight, growth traits also include many morphological traits that have varying degrees of correlation with body weight. Indirect selection of body weight can be achieved through the selection of morphological traits. Therefore, it is necessary to understand the degree of correlation between various morphological traits and body weight,

Table 3.36 Average body weight of grass carp (Chen 1991) (unit: g)

	Age							
Method	1	2	3	4	5	6	7	8
Actual measurement	475	1515	3305	4915	6235	6910	8170	9475
Growth model I	436	1543	3047	4638	6119	7400	8455	2989
Growth model II	446	1892	3281	4615	5896	7126	8308	9443
Growth model III	468	1477	3534	5666	6770	7148	7254	7283

which is an important basis for breeding programs. In this study, Liu et al. (2017) applied the gray relational analysis method to analyze the relationship between morphological traits and body weight of small yellow croaker and analyzed the relative importance of different morphological traits to body weight, which provided important guidance for the development of improved breeding programs for small yellow croaker.

The experimental fish were small yellow croaker populations aged 4.5 months and were cultured in Xixuan Fishery Science and Technology Island of Zhejiang Institute of Marine Fisheries. A total of 123 tails were randomly sampled, with a body weight of 18.172 ± 5.370 g. The morphological traits were accurately measured tail by tail using a Vernier caliper, including full length, body length, head length, trunk length, tail length, tail peduncle length, tail peduncle height, and body height. The statistics were divided into males and females. Because the dimensions of body weight and morphological traits are different, it is impossible to directly compare the traits. Therefore, it is necessary to perform appropriate data conversion. In the study, the standard deviation method was used to dimensionlessly process the data of each trait, and then the data obtained after the transformation were used. Gray correlation analysis was performed. The eight morphological traits and body weight indicators of small yellow croaker were treated as a gray system, the general gray correlation method was used for calculation, and the resolution coefficient was set to 0.5. The degree of association between morphological traits and body weight can be determined according to the degree of gray correlation, thereby determining the relative importance of the morphological traits to body weight.

According to statistics, among all traits, the coefficient of variation of body weight was the largest, and the coefficient of variation between morphological traits was not much different. The minimum, maximum, and mean values of the female samples of small yellow croaker were greater than the corresponding parameters of the male samples. The results of the significance test showed that. The body length and torso length of the female samples were significantly larger than those of the male samples ($P < 0.01$), and the head length, tail stalk length, tail stalk height, and body weight of the male samples were significantly larger than those of the male samples ($P < 0.05$). The difference between male and female samples was not significant ($P > 0.05$), indicating that there was a difference in growth rate between male and female individuals of small yellow croaker, but not all morphological traits showed significant differences.

The calculation results of the gray correlation are shown in Table 3.37. Table 3.37 shows that the gray correlation degree between different morphological traits and body weight of small yellow croaker is in the range of 0.5266–0.6812 (female) and 0.5288–0.7116 (male). The gray correlations between various morphological traits and body weight in male samples were greater than those in female samples. In addition to the tail length trait, the standard deviation of the correlation coefficient of the male samples was also slightly smaller than that of the female samples. The results of the gray correlation between morphological traits and body weight in the female and male samples were as follows: full length > body length > trunk length > tail length > tail stalk length > body height > tail stalk height > head

Table 3.37 Gray correlation between morphological traits and body weight of samples (Liu et al. 2017)

Traits	Female		Male	
	Gray correlation (±SD)	Ranking	Gray correlation (±SD)	Ranking
Full length	0.6812 ± 0.0977	1	0.7116 ± 0.0970	1
Body length	0.6539 ± 0.0956	2	0.6728 ± 0.0905	2
Head length	0.5266 ± 0.0693	8	0.5284 ± 0.0676	8
Body length	0.5658 ± 0.0796	3	0.5729 ± 0.0786	4
Tail length	0.5648 ± 0.0775	4	0.5732 ± 0.0779	3
Caudal peduncle length	0.5424 ± 0.0736	5	0.5472 ± 0.0727	5
Tail stalk height	0.5343 ± 0.0737	7	0.5385 ± 0.0730	6
Body height	0.5345 ± 0.0714	6	0.5378 ± 0.0700	7

length (female); Length > tail length > trunk length > tail stalk length > tail stalk height > body height > head length (male). Therefore, there was a certain difference in the degree of association between the morphological traits and body weight of the female and male samples of small yellow croaker. Therefore, if the data of male and female samples are analyzed separately, the results will be more accurate and reliable. A comprehensive comparison of morphological traits showed that the gray correlation between full length and body weight was the highest (0.6812 for females and 0.7116 for males), followed by body length (0.653 9 for females and 0.672 8 for males). The head length was the lowest (0.5266 for females and 0.5284 for males), and there were certain differences in the ranking of other traits between female and male samples.

In this study, the gray correlation analysis method was used to analyze the correlation between the eight morphological traits of small yellow croaker and body weight. The results showed that the correlation between the full length and the body weight of the female and male samples was the largest, followed by the body length. The head length was the smallest. According to the principle of gray correlation analysis, morphological traits with high correlation are closely related to body weight and vice versa. The effect of head length is relatively small. The correlation between several other morphological traits and body weight showed certain differences between male and female samples. This indicates that when analyzing the growth data of small yellow croaker, it is necessary to analyze the female and male samples separately.

References

Chen CQ (1991) Fishery application of gray relational analysis. J Hydroecol 5:25–27. (In Chinese)

Chen XJ (2003) Application of gray system theory in fishery science. China Agricultural Press. (In Chinese)

Chen XJ (2023) Application of gray system theory in fishery science. China Agricultural Press. (In Chinese)

Chen XJ, Zhou YQ (2002a) Assessment of sustainable use of fisheries resources based on the methods of gray relative relationship. J Fish China 26(4):331–336. (In Chinese)

Chen XJ, Zhou YQ (2002b) Gray relationship analysis of Chinese fishery yield construction. Chinese Fish Econ 2:30–33. (In Chinese)

Chen XJ, Xu LX, Tian SQ (2003) Spatial and temporal analysis of *Ommastrephe bartrami* resources and its fishing ground in North Pacific Ocean. J Fish China 27(4):334–342. (In Chinese)

Deng JL (1987) Basic method of the gray system. Huazhong Technology College Press. (In Chinese)

Deng JL (1990) A course in gray system theory. Huazhong Technology University Press. (In Chinese)

He ZJ (1989) Application of gray system theory to pot fishing experimenting and analyzation. J Zhejiang Coll Fish 8(2):123–126. (In Chinese)

Liu SF, Yang YJ, Wu LF et al (2014) Gray system theory and its application. Science Press, Beijing. (In Chinese)

Liu F, Lou B, Chen RY et al (2017) Analysis of gray relationship between morphological traits and body weight in the small yellow croaker (*Pseudosciaena polyactis*). J Shang Ocean Univ 26(1): 131–137. (In Chinese)

Wang SB (1996) The gray system relevant analysis of the fish catch and its relevant factors of Danjiangkou reservoir. Syst Sci Comprehen Stud Agric 12(1):4–7. (In Chinese)

Wang W (2015) Application of gray correlation analysis on water quality assessment of the *Gymnocypris przewalskii* in aquatic germplasm reserve. Heilongjiang Agric Sci 3:98–100. (In Chinese)

Yan XM, Hu SK, Shi XK (1996) A study on the factors affecting icefish resource and the forecasting of the resources. J Fish China 20(4):307–313. (In Chinese)

Chapter 4
Gray Cluster Analysis

Xinjun Chen

Abstract Clustering is the process of dividing a collection of physical or abstract objects into classes composed of similar objects. In other words, according to a certain standard (such as distance), a data set is divided into different classes or clusters, so that the similarity of data objects within the same cluster is as large as possible, at the same time no longer within the same cluster of data objects as large as possible differences. The cluster generated by clustering is a set of data objects, which are similar to each other and different from other objects in the same cluster. There are many classification problems in natural science and social science. Cluster analysis, also known as group analysis, is a statistical analysis method for the study of classification problems (samples or indicators). Cluster analysis originated from taxonomy, but cluster is not equal to classification. The difference between clustering and classification is that the classification is unknown, clustering is an unsupervised learning task, do not know the real sample markers, only make the similarity of samples together. Classification is a supervised learning task, which uses known sample markers to train the learner to predict the classification of unknown samples. Clustering can be used as a separate process to find the inherent distribution structure of data, and it can also be used as a precursor process of other learning tasks. The content of clustering analysis is very rich, including systematic clustering, ordered samples clustering, dynamic clustering, fuzzy clustering, graph theory clustering, gray clustering, and so on. Clustering is also an important concept in data mining. Gray clustering is a method to aggregate some observation indexes or objects into several definable categories according to the correlation matrix or the whitening weight function of gray number, it can be divided into gray constellation clustering, gray relational clustering, and gray whitening function clustering. In this chapter, the concept and types of gray clustering are introduced, and the principles and calculation methods of gray constellation clustering, gray relational clustering, gray variable weight clustering, and gray fixed-weight clustering are expounded, at the same time, the application of gray clustering in fishery science is analyzed.

X. Chen (✉)
College of Marine Sciences, Shanghai Ocean University, Lingang New City, Shanghai, China
e-mail: xjchen@shou.edu.cn

X. Chen (ed.), *Application of Gray System Theory in Fishery Science*,
https://doi.org/10.1007/978-981-99-0635-2_4

Keywords Gray clustering · Gray constellation clustering · Gray relational clustering · Gray variable weight clustering · Gray fixed-weight clustering

4.1 Concept of Gray Clustering

Gray clustering is a method that aggregates some observation indicators or observation objects into several definable categories according to the correlation matrix or the whitening weight function of gray numbers. A cluster can be viewed as a collection of objects belonging to the same type of observation. In practical problems, each observation object often has many characteristic indicators, which makes it difficult to perform accurate classification.

According to different clustering methods, gray clustering can be divided into gray constellation clustering, gray relational clustering, and gray-type whitening function clustering. Gray constellation clustering quantitatively determines the relationship between samples based on their own attributes, uses the principle of similarity, and performs natural clustering based on this relationship. Gray relational clustering is mainly used for the merging of similar factors to simplify complex systems. Through gray relational clustering, we can analyze whether there are several factors in many factors that are closely related so that the comprehensive average index of these factors or one of them can be used to represent these factors in the study, and at the same time, the information is not available. Therefore, before conducting a large-scale survey, the collection of unnecessary variables (factors) can be reduced through the gray correlation clustering of typical sampling data to save costs and funds. Gray class whitening weight function clustering is mainly used to check whether the observed objects belong to different preset categories so that they can be treated differently. In terms of computational workload, gray-type whitening functions are more complex than gray correlation clustering and constellation clustering (Chen 2003, 2023).

4.2 Gray Constellation Clustering

4.2.1 *Principles and Methods*

Constellation clustering is a relatively simple and easy-to-implement clustering method in gray clustering. The basic principle is as follows: each sample point is placed in an upper semicircle according to a certain number of relationships, a sample point is represented by a "star point," and the sample points of the same type form a "constellation," which is then sketched. The boundary between different constellations is drawn so that classification can be carried out. In essence, it transforms a large amount of information (or index value) in a sample into dimensionless by means of transformation (range transformation) of the original data and becomes a simple problem of spatial coordinate comparison.

In general, constellation clustering has the following steps:

1. Range transformation is performed on the original index values, and the transformed values fall within the closed interval of [0°, 180°]. The calculation formula is

$$a_{ij} = \frac{X_{ij} - X_{j\min}}{X_{j\max} - X_{j\min}} \times 180^\circ$$

where a_{ij} is the transformed data, expressed as an angle; X_{ij} is the raw data; $X_{j\max}$ is the maximum value of the jth variable; $X_{j\min}$ is the minimum value of the jth variable; ($j = 1, 2, 3 \ldots$, P is the number of indicators); ($i = 1, 2, 3 \ldots$, N is the sample number).
2. For each indicator, a weight W_j is given according to its degree of influence on the system change, so that

$$\sum_{j=1}^{p} w_j = 1$$

$$0 < w_{j<1.}$$

where w_j is the weight of the jth index
3. Calculate the Cartesian coordinates of each indicator.

Using the transformation relationship between polar coordinates and rectangular coordinates, the values of X_i and Y_i of each index of each point are first obtained, and then the values of X_i and Y_i of each index of each point are added together to obtain the coordinates of each sample point. Value. Its transformation formula is

$$X_i = \sum_{j=1}^{p} W_j \cos a_{ij}$$

$$Y_i = \sum_{j=1}^{p} W_j \sin a_{ij}$$

where X_i is the abscissa of the ith sample point; Y_i is the ordinate of the ith sample point.
4. Draw a constellation diagram

Draw an upper semicircle with a radius of 1, using the center of the circle as the origin of coordinates and the base of the upper semicircle as the X-axis. Similar and close sample points are clustered together to form a "constellation."
5. Calculate the comprehensive index value

The mathematical expression of the comprehensive index value is

$$Z_i = \sum_{j=1}^{p} a_{ij} W_j \; 0 < Z_i < 1$$

where Z_i is the comprehensive index value.

6. Cluster analysis

According to the comprehensive index value and the clustering situation of the constellation diagram, the classification result is determined.

4.2.2 Example Analysis

It is assumed that there are seven samples (Table 4.1), and each sample has six indicators: X_0, X_1, X_2, X_3, X_4, and X_5. The seven samples were classified using the constellation clustering method.

Step 1: Calculate the extreme value of each variable

$$X_{0\max} = 2.73 \; X_{0\min} = 2.54$$

$$X_{1\max} = 63.19 \; X_{1\min} = 38.60$$

$$X_{2\max} = 67.70 \; X_{2\min} = 41.10$$

$$X_{3\max} = 69.78 \; X_{3\min} = 24.84$$

$$X_{4\max} = 2.61 \; X_{4\min} = 1.26$$

$$X_{5\max} = 1.18 \; X_{5\min} = 0.88$$

Step 2: Range transformation

Based on the formula $a_{ij} = \frac{X_{ij} - X_{j\min}}{X_{j\max} - X_{j\min}} \times 180°$, the transformation data are obtained:

$$a_{01} = \frac{X_{01} - X_{0\min}}{X_{0\max} - X_{0\min}} \times 180° = \frac{2.64 - 2.54}{2.73 - 2.54} \times 180 = 94.74°$$

By analogy, the data in Table 4.2 can bc obtained.

Table 4.1 Original data table (Chen 2003, 2023)

Sample	1	2	3	4	5	6	7
X_0	2.64	2.72	2.73	2.72	2.64	2.63	2.54
X_1	63.19	59.12	46.48	51.06	48.18	38.6	41.03
X_2	43.6	41.1	56.9	58.5	62.2	64.5	67.7
X_3	69.78	59.44	51.05	43.15	36.68	29.15	24.84
X_4	2.61	2.24	1.55	1.48	1.44	1.30	1.26
X_5	1.18	1.05	1.04	0.96	0.94	0.88	0.89

Table 4.2 Range transformation data (Chen 2003, 2023)

Sample	1	2	3	4	5	6	7
α_0	94.74	170.53	180.00	170.53	94.74	85.26	0.00
α_1	180.00	150.21	57.68	91.21	70.13	0.00	17.79
α_2	16.92	0.00	106.92	117.74	142.78	158.35	180.00
α_3	180.00	138.58	104.98	73.34	47.42	17.26	0.00
α_4	180.00	130.67	38.67	29.33	24.00	5.33	0.00
α_5	180.00	102.00	96.00	48.00	36.00	0.00	6.00

Step 3: Take weights and perform Cartesian coordinate transformation on each indicator.

It is assumed that the weights of all indicators are equal, i.e., 1/6.

Based on the formula $X_i = \sum_{j=1}^{p} W_j \cos a_{ij}$, $Y_i = \sum_{j=1}^{p} W_j \sin a_{ij}$, the following data are obtained:

$$X_1 = \sum_{j=0}^{6} W_j \cos a_{1j} = \frac{1}{6} \times (\cos 94.74 + \cos 180 + \ldots + \cos 180) = -0.52$$

$$Y_1 = \sum_{j=0}^{6} W_j \sin a_{1j} = \frac{1}{6} \times (\sin 94.74 + \sin 180 + \ldots + \sin 180) = 0.21$$

The rest are similar.

$$X_2 = -0.41, \quad Y_2 = 0.51$$

$$X_3 = -0.06, \quad Y_3 = 0.73$$

$$X_4 = 0.06, \quad Y_4 = 0.71$$

$$X_5 = 0.31, \quad Y_5 = 0.71$$

$$X_6 = 0.52, Y_6 = 0.29$$

$$X_7 = 0.66, \quad Y_2 = 0.07$$

Step 4: Draw a constellation diagram (Fig. 4.1)

Step 5: Calculate the comprehensive index value

From formula $Z,_i = \sum_{j=1}^{p} a_{ij} W_j$, the following is obtained:

$$Z_1 = \sum_{j=0}^{5} a_{1j} W_j = \frac{1}{6} \times (94.74 + 180 + \ldots + 180) = 138.61$$

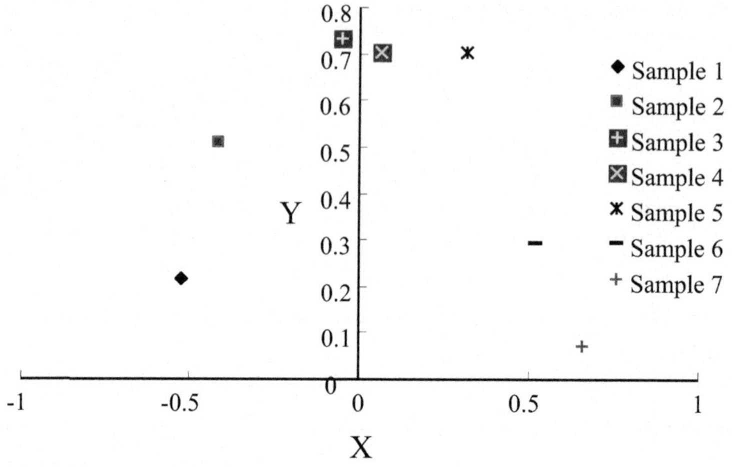

Fig. 4.1 Constellation cluster map (Chen 2003, 2023)

$$Z_2 = 115.33$$

$$Z_3 = 97.37$$

$$Z_4 = 88.36$$

$$Z_5 = 69.18$$

$$Z_6 = 44.37$$

$$Z_7 = 33.96$$

Step 6: Cluster analysis

According to the comprehensive index value and the clustering situation of the constellation diagram, we believe that it can be divided into three categories: sample 1 and sample 2 are one category; sample 3, sample 4, and sample 5 are one category; and sample 6 and sample 7 are one category. The comprehensive evaluation values were 126.97, 84.97, and 39.17, respectively, indicating that the differences between the three categories were extremely significant.

4.3 Gray Relational Clustering

4.3.1 The Basic Method of Gray Relational Clustering

Gray correlation clustering actually uses the basic principle of gray correlation to calculate the degree of correlation between samples and then classifies the types of samples according to the degree of gray correlation. The calculation principle and method are as follows:

Now, there are m samples, each sample has n indicators, and the following sequence is obtained:

$$X_1 = (x_1\ (1), x_1\ (2),\ \ldots, x_1\ (n))$$
$$X_2 = (x_2\ (1), x_2\ (2),\ \ldots, x_2\ (n))$$
$$\cdots\cdots\cdots\cdots\cdots\cdots\cdots\cdots\cdots\cdots$$
$$X_m = (x_m\ (1), x_m\ (2),\ \ldots, x_m\ (n))$$

For all $i \leq j$, $i, j = 1, 2, \ldots, m$, calculate the gray correlation degree ε_{ij} between X_i and X_j, thus obtaining the upper triangular matrix A.

$$A = \begin{bmatrix} \varepsilon_{11} & \varepsilon_{12} & \cdots & \varepsilon_{1m} \\ & \varepsilon_{22} & \cdots & \varepsilon_{2m} \\ & & \ddots & \vdots \\ & & & \varepsilon_{mm} \end{bmatrix}, \text{ where } \varepsilon_{ii} = 1;\ i = 1, 2, \ldots, m;$$

If we take the critical value $r \in [0, 1]$, it is generally required that $r > 0.5$. When $\varepsilon_{ij} \geq r$, X_i and X_j can be regarded as the same type of features.

r can be determined according to the needs of the actual problem. If r is closer to 1, then the classification is finer, and there are fewer variables in each group. In contrast, the classification is more, and there are more variables in each group.

4.3.2 *Case Analysis*

We continue to analyze the examples in the previous section and use the gray absolute correlation calculation method (Chap. 3) for cluster analysis. In this example, there are a total of 7 samples, and each sample has 6 indicators. To save the cost of future surveys and the collection of sample data, we need to classify the indicators to streamline the indicators.

Step 1: Zeroing the starting point

Using $X_i^0(k) = x_i(k) - x_i(1)$, the initialized data in Table 4.3 can be obtained:

Step 2: Seeking $|s_0|$, $|s_i|$ and $|s_i - s_j|$

$$|s_0| = 0.19$$
$$|s_1| = 83.59$$
$$|s_2| = 77.25$$
$$|s_3| = 151.92$$

Table 4.3 Initialized data (Chen 2003, 2023)

Serial number	Sample 1	Sample 2	Sample 3	Sample 4	Sample 5	Sample 6	Sample 7
X^0_0	0.00	0.08	0.09	0.08	0.00	−0.01	−0.10
X^0_1	0.00	−4.07	−16.71	−12.13	−15.01	−24.59	−22.16
X^0_2	0.00	−2.50	13.30	14.90	18.60	20.90	24.10
X^0_3	0.00	−10.34	−18.74	−26.63	−33.10	−40.63	−44.95
X^0_4	0.00	−0.37	−1.06	−1.13	−1.17	−1.31	−1.35
X^0_5	0.00	−0.13	−0.14	−0.22	−0.24	−0.30	−0.29

$$|s_4| = 5.72$$

$$|s_5| = 1.18$$

$$|s_1 - s_0| = 83.78$$

$$|s_2 - s_0| = 77.06$$

$$|s_3 - s_0| = 152.11$$

$$|s_4 - s_0| = 5.91$$

$$|s_5 - s_0| = 1.37$$

Step 3: Find the degree of gray absolute correlation

$$\varepsilon_{01} = 0.50$$

$$\varepsilon_{02} = 0.50$$

$$\varepsilon_{03} = 0.50$$

$$\varepsilon_{04} = 0.54$$

$$\varepsilon_{05} = 0.63$$

$$\varepsilon_{12} = 0.50$$

$$\varepsilon_{13} = 0.78$$

$$\varepsilon_{14} = 0.54$$

$$\varepsilon_{15} = 0.51$$

$$\varepsilon_{23} = 0.49$$

$$\varepsilon_{24} = 0.47$$

$$\varepsilon_{25} = 0.47$$

$$\varepsilon_{34} = 0.52$$

$$\varepsilon_{35} = 0.51;$$

$$\varepsilon_{45} = 0.64$$

Then, the absolute matrix A can be obtained.

$$A = \begin{vmatrix} 1 & 0.50 & 0.50 & 0.50 & 0.54 & 0.63 \\ & 1 & 0.50 & 0.78 & 0.54 & 0.51 \\ & & 1 & 0.49 & 0.47 & 0.47 \\ & & & 1 & 0.52 & 0.51 \\ & & & & 1 & 0.64 \\ & & & & & 1 \end{vmatrix}$$

If we assume that the critical value of the absolute correlation degree is 0.60, then we can check in turn X_5 and X_0, X_3 and X_1, and X_5 and X_4. Taking the indicator with the smallest label as the representative of each category, X_5 and X_4 can be combined into X_0 to form a category so that the clustering results of the six indicators are

$$\{X_5, X_4, X_0\}, \{X_3, X_1\}, \{X_2\}$$

In other words, in the future collection of sample data, we only need to collect the data of three indicators, X_0, X_1, and X_2.

4.4 Gray Variable Weight Clustering

4.4.1 Concept and Method of Gray Variable Weight Clustering

Suppose there are n clustering objects, m clustering indicators, and s different gray classes. According to the ith ($i = 1, 2, \ldots, n$) object with respect to the j ($j = 1, 2, \ldots, m$) index, the sample value x_{ij} ($i = 1, 2, \ldots, n; j = 1, 2, \ldots, m$) classifies the ith object into the k ($k \in \{1, 2, \ldots, s\}$) gray class. Among them, it is called gray clustering.

Assume now that the whitening weight function of j index k subclasses $f_j^k(\cdot)$ is the typical whitening weight function shown in Fig. 4.2, $x_j^k(1), x_j^k(2), x_j^k(3), x_j^k(4)$ are the turning point of $f_j^k(\cdot)$. The typical whitening weight function is denoted as (Liu et al. 2014):

$$f_j^k\left[x_j^k(1), x_j^k(2), x_j^k(3), x_j^k(4)\right]$$

Fig. 4.2

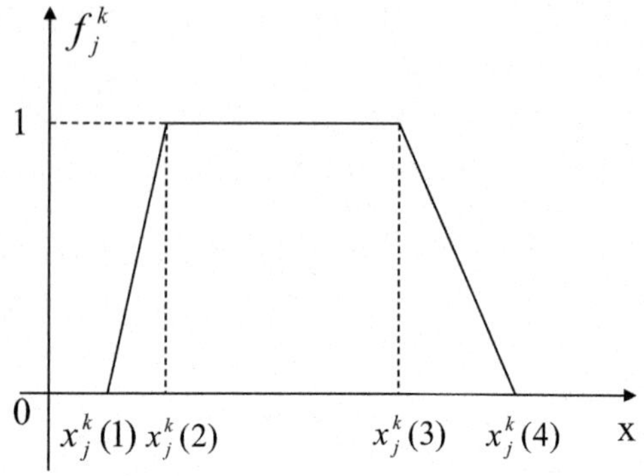

Fig. 4.3

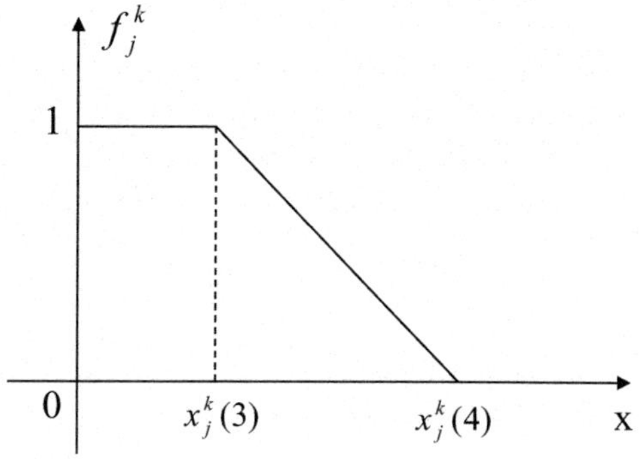

Fig. 4.4

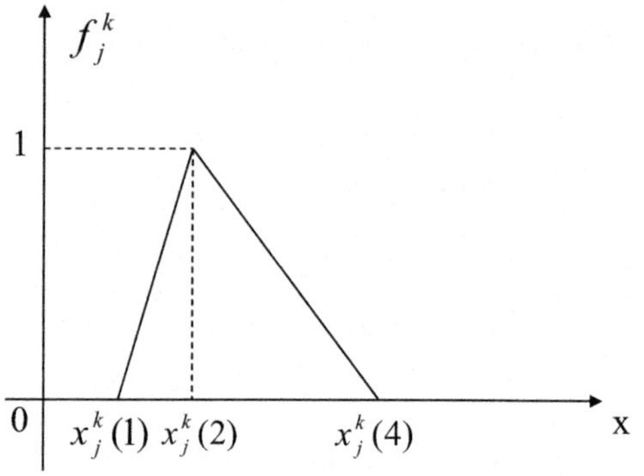

If whitening weight function $f_j^k(\cdot)$ has no first turning point $x_j^k(1)$ and the second turning point $x_j^k(2)$, as shown in Fig. 4.3, then $f_j^k(\cdot)$ is called the lower limit measure of the whitening weight function, denoted as $f_j^k\left[-,\,-,x_j^k(3),x_j^k(4)\right]$.

If the whitening weight function $f_j^k(\cdot)$ coincides between the second turning point of $x_j^k(2)$ and the third turning point $x_j^k(3)$, as shown in Fig. 4.4, then $f_j^k(\cdot)$ is called a

Fig. 4.5

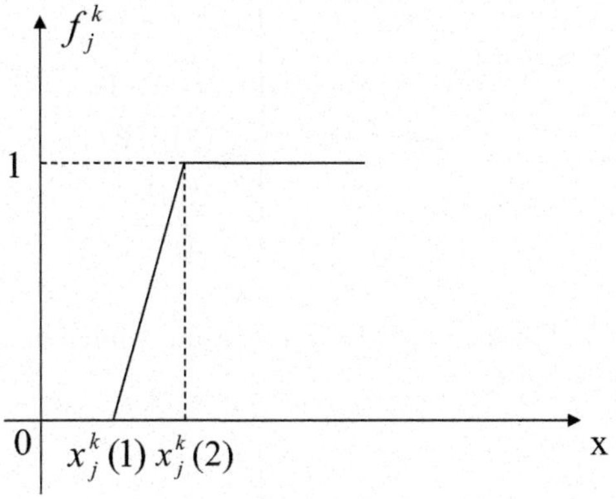

moderate measure of the whitening weight function, denoted as $f_j^k\left[x_j^k(1), x_j^k(2), -, x_j^k(4)\right]$.

If whitening weight function $f_j^k(\cdot)$ has no third turning point $x_j^k(3)$ and the fourth turning point $x_j^k(4)$, as shown in Fig. 4.5, then $f_j^k(\cdot)$ is called the upper limit measure whitening weight function, denoted as $f_j^k\left[x_j^k(1), x_j^k(2), -, -\right]$.

Through the above analysis, we can obtain the whitening weight function under different conditions.

1. For the typical whitening function shown in Fig. 4.2,

$$f_j^k(x) = \begin{cases} 0 & x \notin \left[x_j^k(1), x_j^k(4)\right] \\ \dfrac{x - x_j^k(1)}{x_j^k(2) - x_j^k(1)} & x \in \left[x_j^k(1), x_j^k(2)\right] \\ 1 & x \in \left[x_j^k(2), x_j^k(3)\right] \\ \dfrac{x_j^k(4) - x}{x_j^k(4) - x_j^k(3)} & x \in \left[x_j^k(3), x_j^k(4)\right] \end{cases}$$

2. For the lower bound measure whitening weight function shown in Fig. 4.3,

$$f_j^k(x) = \begin{cases} 0 & x \notin \left[0, x_j^k(4)\right] \\ 1 & x \in \left[0, x_j^k(3)\right] \\ \dfrac{x_j^k(4) - x}{x_j^k(4) - x_j^k(3)} & x \in \left[x_j^k(3), x_j^k(4)\right] \end{cases}$$

3. For the whitening weight function of the moderate measure shown in Fig. 4.4, there is

$$f_j^k(x)=\begin{cases}0 & x\notin\left[x_j^k(1),x_j^k(4)\right]\\ \dfrac{x-x_j^k(1)}{x_j^k(2)-x_j^k(1)} & x\in\left[x_j^k(1),x_j^k(2)\right]\\ \dfrac{x_j^k(4)-x}{x_j^k(4)-x_j^k(2)} & x\in\left[x_j^k(2),x_j^k(4)\right]\end{cases}$$

4. For the upper bound measure whitening weight function shown in Fig. 4.5, we have

$$f_j^k(x)=\begin{cases}0 & x\prec x_j^k(1)\\ \dfrac{x-x_j^k(1)}{x_j^k(2)-x_j^k(1)} & x\in\left[x_j^k(1),x_j^k(2)\right]\\ 1 & x\geq x_j^k(2)\end{cases}$$

For the whitening weight function of the *j*th index *k* subcategory shown in Fig. 4.2, let $\lambda_j^k=\frac{1}{2}\left(x_j^k(2)+x_j^k(3)\right)$.

For the whitening weight function of the *jth* index *k* subcategory shown in Fig. 4.3, let $\lambda_j^k=x_j^k(3)$.

For the whitening weight function of the *j*th index *k* subcategory shown in Fig. 4.4 and Fig. 4.5, let $\lambda_j^k=x_j^k(2)$;

We call λ_j^k the critical value of subcategory *k* of index *j* and call $\eta_j^k=\frac{\lambda_j^k}{\sum_{j=1}^m\lambda_j^k}$ the weight of index *j* with respect to the *k* subcategories.

Now let x_{ij} be the specimen of object *i* with respect to index *j*, $f_j^k(\cdot)$ is the whitening weight function of *j* index *k* subclass, η_j^k is the weight of the *j* index with respect to the *k* subcategory, then $\sigma_i^k=\sum_{j=1}^m f_j^k(x_{ij})\times\eta_j^k$ is called the gray variable weight clustering coefficient for the object *i* belonging to *k* gray class.

Let $\sigma_i=\left(\sigma_i^1,\sigma_i^2,\cdots,\sigma_i^s\right)=\left(\sum_{j=1}^m f_j^1(x_{ij})\times\eta_j^1,\sum_{j=1}^m f_j^2(x_{ij}),\cdots,\sum_{j=1}^m f_j^s(x_{ij})\times\eta_j^s\right)$ is the clustering coefficient vector of object *i*.

Now, set $\sigma_i^{k^*}=\max_{1\leq k\leq s}\left\{\sigma_i^k\right\}$; then, it is said that object *i* belongs to gray class k^*.

Gray variable weight clustering is applicable to situations in which the meaning and dimension of the indicators are the same. When the significance and dimension of the clustering indicators are different and the sample values of different indicators are very different in number, it is not appropriate to use gray variable weight clustering.

4.4.2 Example Analysis

There are three coastal fishing areas, and the three clustering indicators are the output value of marine fishing, the output value of aquaculture, and the output value of aquatic product processing. The sample data are shown in matrix A:

$$A=\left(x_{ij}\right)=\begin{bmatrix} x_{11} & x_{12} & x_{13} \\ x_{21} & x_{22} & x_{23} \\ x_{31} & x_{32} & x_{33} \end{bmatrix}=\begin{bmatrix} 80 & 20 & 100 \\ 40 & 30 & 30 \\ 10 & 90 & 60 \end{bmatrix}$$

Clustering was performed based on high, medium, and low output values.

It is now assumed that the whitening weight functions for the indicators of marine fishing output, aquaculture output, and aquatic product processing output are

$$f_1^1[0,80,-,-],f_1^2[0,40,-,80],f_1^3[-,-,10,20]$$
$$f_2^1[0,90,-,-],f_2^2[0,45,-,90],f_2^3[-,-,15,30]$$
$$f_3^1[0,100,-,-],f_3^2[0,50,-,100],f_3^3[-,-,20,40]$$

From the above whitening weight functions, they are

$$f_1^1(x)=\begin{cases} 0, & x<0 \\ \dfrac{x}{80}, & 0\le x<80 \\ 1, & x>80 \end{cases} \quad f_1^2(x)=\begin{cases} 0, & x<0 \\ \dfrac{x}{40}, & 0\le x\le 40 \\ \dfrac{80-x}{40}, & 40<x\le 80 \\ 0, & x>80 \end{cases}$$

$$f_1^3(x)=\begin{cases} 0, & x<0 \\ 1, & 0\le x\le 10 \\ \dfrac{20-x}{10}, & 10<x\le 20 \\ 0, & x>20 \end{cases} \quad f_2^1(x)=\begin{cases} 0, & x<0 \\ \dfrac{x}{90}, & 0\le x<90 \\ 1, & x>90 \end{cases}$$

$$f_2^2(x)=\begin{cases}0 & , \quad x<0\\ \dfrac{x}{45} & , \quad 0\le x\le 45\\ \dfrac{90-x}{45} & , \quad 45<x\le 90\\ 0 & , \quad x>90\end{cases} \qquad f_2^3(x)=\begin{cases}0 & , \quad x<0\\ 1 & , \quad 0\le x\le 15\\ \dfrac{30-x}{15} & , \quad 15<x\le 30\\ 0 & , \quad x>30\end{cases}$$

$$f_3^1(x)=\begin{cases}0 & , \quad x<0\\ \dfrac{x}{100} & , \quad 0\le x<100\\ 1 & , \quad x>100\end{cases} \qquad f_3^2(x)=\begin{cases}0 & , \quad x<0\\ \dfrac{x}{50} & , \quad 0\le x\le 50\\ \dfrac{100-x}{50} & , \quad 50<x\le 100\\ 0 & , \quad x>100\end{cases}$$

$$f_3^3(x)=\begin{cases}0 & , \quad x<0\\ 1 & , \quad 0\le x\le 20\\ \dfrac{40-x}{20} & , \quad 20<x\le 40\\ 0 & , \quad x>40\end{cases}$$

Thus, we have

$$\lambda_1^1=80,\ \lambda_2^1=90,\ \lambda_3^1=100$$

$$\lambda_1^2=40,\ \lambda_2^2=45,\ \lambda_3^2=50$$

$$\lambda_1^3=10,\ \lambda_2^3=15,\ \lambda_3^3=20$$

From the formula $\eta_j^k=\lambda_j^k/\sum_{j=1}^{3}\lambda_j^k$ obtained:

$$\eta_1^1=\frac{80}{80+90+100}=\frac{80}{270}$$

$$\eta_2^1=\frac{90}{80+90+100}=\frac{90}{270}$$

$$\eta_3^1=\frac{100}{80+90+100}=\frac{100}{270}$$

$$\eta_1^2=\frac{40}{135}$$

$$\eta_2^2 = \frac{45}{135}$$

$$\eta_3^2 = \frac{50}{135}$$

$$\eta_1^3 = \frac{10}{45}$$

$$\eta_2^2 = \frac{15}{45}$$

$$\eta_3^2 = \frac{20}{45}$$

Then, when $i = 1$, from the formula $\sigma_i^k = \sum_{j=1}^{m} f_j^k(x_{ij}) \times \eta_j^k$ we have

$$\delta_1^1 = \sum_{j=1}^{m} f_j^k(x_{ij}) \times \eta_j^k = f_1^1(x_{11}) \times \eta_1^1 + f_2^1(x_{12}) \times \eta_2^1 + f_3^1(x_{13}) \times \eta_3^1$$
$$= f_1^1(80) \times \frac{80}{270} + f_2^1(20) \times \frac{90}{270} + f_3^1(100) \times \frac{100}{270}$$
$$= 1 \times \frac{80}{270} + \frac{20}{90} \times \frac{90}{270} + 1 \times \frac{100}{270} = 0.74$$

Similarly, we can obtain:

$$\sigma_1^2 = 0.15$$

$$\sigma_1^3 = 0.22$$

$$\sigma_2^1 = 0.37$$

$$\sigma_2^2 = 0.74$$

$$\sigma_2^3 = 0.22$$

$$\sigma_3^1 = 0.59$$

$$\sigma_3^2 = 0.15$$

$$\sigma_3^3 = 0.22$$

Then, the following formula $\sigma_i^{k^*} = \max_{1 \le k \le s} \{\sigma_i^k\}$ is obtained:

$$\max_{1 \le k \le 3} \{\sigma_1^k\} = \max\{\delta_1^1, \delta_1^2, \delta_1^3\} = \max\{0.74, 0.15, 0.22\} = \delta_1^1$$

$$\max_{1 \le k \le 3}\{\sigma_2^k\} = \max\{\delta_2^1, \delta_2^2, \delta_2^3\} = \max\{0.37, 0.74, 0.22\} = \delta_2^2$$

$$\max_{1 \le k \le 3}\{\sigma_3^k\} = \max\{\delta_3^1, \delta_3^2, \delta_3^3\} = \max\{0.59, 0.15, 0.22\} = \delta_3^1$$

The above results indicate that the second fishing area is a moderately developed area of the fishery economy, and the first and third fishing areas are highly developed areas.

4.5 Gray Fixed-Weight Clustering

4.5.1 General Method of Gray Fixed-Weight Clustering

When the clustering indicators have different meanings and dimensions and there is a large disparity in the number, if we adopt gray variable weight clustering, the effect of some indicators in the clustering may be very weak. There are two methods to solve this problem: one is to use the raw data processing methods in Chap. 2 (such as the initial value or average) for dimensionless processing and then perform clustering. This method treats all clustering indicators equally and cannot reflect the difference in the role of different indicators in the clustering process. Another method is to assign weights to each clustering index in advance. There are many methods to assign weights, and the analytic hierarchy process (AHP) is generally used. This method is a clustering method that assigns weights in advance, so it is called gray fixed-weight clustering.

Gray fixed-weight clustering can be performed according to the following steps:

Step 1: Give the whitening weight function of the *j*th index *k* subcategory $f_j^k(\cdot)$ ($J = 1, 2, \ldots, m$; $k = 1, 2, \ldots, s$).

Step 2: Determine the clustering weight η_j ($j = 1, 2..., m$) of each index according to the qualitative analysis conclusion.

Step 3: Use the whitening weight function obtained in Steps 1 and 2 $f_j^k(\cdot)$ ($J = 1, 2, \ldots, m$; $k = 1, 2, \ldots, s$), clustering weight η_j ($j = 1, 2, \ldots, m$), and the sample value x_{ij} of object *i* with respect to *j* index ($i = 1, 2, \ldots, n$; $j = 1, 2, \ldots, m$) to calculate the fixed-weight clustering coefficient $\sigma_i^k = \sum_{j=1}^{m} f_j^k(x_{ij}) \times \eta_j$, $I = 1, 2, \ldots, s$.

Step 4: If $\sigma_i^{k^*} = \max_{1 \le k \le s}\{\sigma_i^k\}$, then it is concluded that object *i* belongs to gray class k^*.

4.5.2 Example Analysis

We also illustrate the examples in the previous section.

Step 1: Determine the whitening weight function as in the previous section.

Step 2: Determine the cluster weights of fishing, aquaculture, and processing indicators.

$$\eta_1 = 0.50,\ \eta_2 = 0.30,\ \eta_3 = 0.20$$

Step 3: From the formula $\sigma_i^k = \sum_{j=1}^{m} f_j^k(x_{ij}) \times \eta_j$, calculate

$$\begin{aligned}\sigma_1^1 &= \sum_{j=1}^{3} f_j^1(x_{1j}) \times \eta_j = f_1^1(x_{11}) \times \eta_1 + f_2^1(x_{12}) \times \eta_2 + f_3^1(x_{13}) \times \eta_3 \\ &= f_1^1(80) \times 0.5 + f_2^1(20) \times 0.3 + f_3^1(100) \times 0.2 \\ &= 1 \times 0.5 + \frac{20}{90} \times 0.3 + 1 \times 0.2 = 0.756\end{aligned}$$

Similarly, we can obtain:

$$\sigma_1^3 = 0.20$$
$$\sigma_2^1 = 0.41$$
$$\sigma_2^2 = 0.82$$
$$\sigma_2^3 = 0.10$$
$$\sigma_3^1 = 0.48$$
$$\sigma_3^2 = 0.29$$
$$\sigma_3^3 = 0.50$$

Then, the following formula $\sigma_i^{k^*} = \max_{1 \le k \le s}\{\sigma_i^k\}$ is obtained:

$$\max_{1 \le k \le 3}\{\sigma_1^k\} = \max\{\delta_1^1, \delta_1^2, \delta_1^3\} = \max\{0.756, 0.13, 0.20\} = \delta_1^1$$

$$\max_{1 \le k \le 3}\{\sigma_2^k\} = \max\{\delta_2^1, \delta_2^2, \delta_2^3\} = \max\{0.41, 0.82, 0.10\} = \delta_2^2$$

$$\max_{1 \le k \le 3}\{\sigma_3^k\} = \max\{\delta_3^1, \delta_3^2, \delta_3^3\} = \max\{0.48, 0.29, 0.50\} = \delta_3^3$$

The above results indicate that the second fishing area belongs to an area with a moderately developed fishery economy, the first one belongs to a highly developed area, and the third fishing area belongs to an underdeveloped area of fisheries.

4.6 Analysis of the Application of Gray Clustering in Fishery Science

Gray clustering has been widely used in fishery science, mainly in the aspects of fishery regional economy, fish population division, fish nutritional value evaluation, and fishery water environment evaluation. The application of gray clustering in fishery science is specifically analyzed based on the reports in the relevant literature.

4.6.1 Application of Gray Clustering in the Fishery Regional Economy

In the study of the fishery regional economy, it is very important to formulate future development strategies based on the natural resource conditions of each region and the development level of the fishery economy. Therefore, it is very important to use the gray clustering method to scientifically classify the fishery economy in China's coastal areas to formulate different types of development plans and implement classification guidance to ensure the sustainable development of the fishery economy in coastal areas. Meaning.

Chen and Zhang (2001) published a preliminary study on the types of fishery economic regions in China's coastal provinces and cities. The statistical data of coastal fishery development in 11 provinces in China in 1997 were selected, namely Tianjin, Hebei, Liaoning, Shanghai, Jiangsu, Zhejiang, Fujian, Shandong, Guangdong, Guangxi, and Hainan. Based on the characteristics and development of the aquaculture industry, a comprehensive technical and economic evaluation index for fisheries was established, and 17 indicators were used. There are the total production of aquatic products (tons), marine fishing production (tons), marine aquaculture production (tons), freshwater fishing yield (ton), freshwater aquaculture yield (ton), offshore fishery yield (ton), total output value (10,000 yuan), aquatic product processing yield (ton), per capita aquatic product yield (ton/person), per capita total fishery output value (yuan/person), per capita net income of fishermen (yuan/person), per capita net income of fishermen (yuan/person), per capita investment in fixed assets (yuan/person), average per unit yield of seawater and freshwater aquaculture (kg/ha), the water surface utilization rate of mariculture (%), and the water surface utilization rate of freshwater aquaculture (%). Gray constellation clustering was used to preliminarily classify the types of fishery economic regions in 11 coastal provinces in China.

In the calculation, the weight of each indicator is set to 1/17, and the values of the abscissa (X_i) and the ordinate (Y_i) of each province are calculated, as shown in Table 4.4. The constellation clustering diagram is plotted in Fig. 4.6. The calculated comprehensive index value (Z) is shown in Tables 4.5.

According to the comprehensive index values of the 11 coastal provinces and the clustering distribution map (Fig. 4.6), fishery economic development can be divided

Table 4.4 Cluster analysis results (Chen and Zhang 2001)

Sample point	Tianjin	Hebei	Liaoning	Shanghai	Jiangsu	Zhejiang
X value	0.5263	0.5896	0.2052	0.4064	0.3354	0.0829
Y value	0.3396	0.4568	0.6318	0.3548	0.4102	0.6794
Sample point	Fujian	Shandong	Guangdong	Guangxi	Hainan	
X value	0.0276	−0.3990	0.0224	0.6367	0.7256	
Y value	0.5765	0.5855	0.5708	0.4812	0.3004	

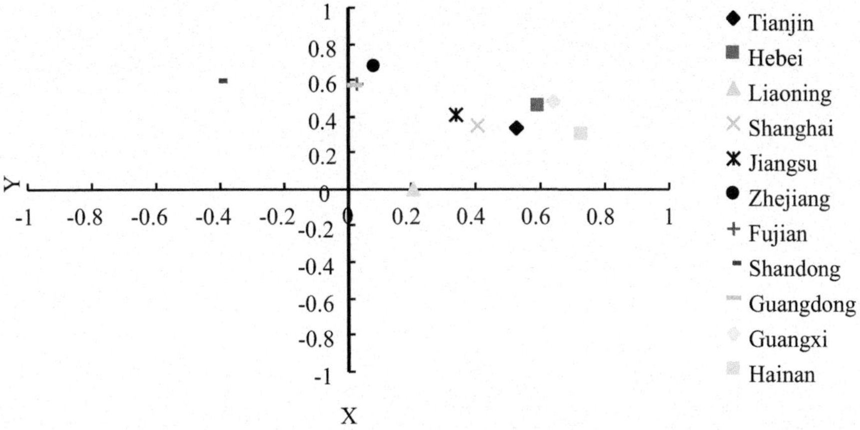

Fig. 4.6 Gray clustering results of comprehensive index values of various provinces and cities (Chen and Zhang 2001)

Table 4.5 Comprehensive indicator values of the fishery economy in various provinces and cities (Chen and Zhang 2001)

Province	Tianjin	Hebei	Liaoning	Shanghai	Jiangsu	Zhejiang
Comprehensive value	0.5263	0.5896	0.2052	0.4064	0.3354	0.0829
Province	Fujian	Shandong	Guangdong	Guangxi	Hainan	
Comprehensive value	0.0276	−0.3990	0.0224	0.6367	0.7256	

into four regions. Among the four provinces, category III includes five provinces and municipalities, namely Jiangsu, Shanghai, Guangxi, Hebei, and Tianjin, and category IV includes Hainan Province (Table 4.6).

4.6.2 Application of Gray Clustering in the Evaluation of the Nutritional Value of Fish

Xie and Pan (1992) published "A Comprehensive Evaluation of the Nutritional Value of Several Famous and High-quality Fishes." The authors used the protein

Table 4.6 Results of the division of fishery economic zones (Chen and Zhang 2001)

Type	Province	Z value	Type	Province	Z value	Type	Province	Z value	Type	Province	Z value
Class I	Shandong	0.684	Class II	Guangdong	0.514	Class III	Jiangsu	0.356	Class IV	Hainan	0.163
			Class II	Fujian	0.496	Class III	Shanghai	0.308			
			Class II	Zhejiang	0.465	Class III	Hebei	0.253			
			Class II	Liaoning	0.441	Class III	Tianjin	0.253			
						Class III	Guangxi	0.241			

Table 4.7 Evaluation criteria for the nutritional value of freshwater fish in China (Xie and Pan 1992)

Evaluation parameters	Low	Medium	High
Protein content (%)	≤10	15	≥20
Fat content (%)	≤1	3	≥10
Essential Amino Acid Index (%)	≤20	50	≥80
Flavored amino acid content (%)	≤10	20	≥30

Table 4.8 Nutritional composition of six fish species (Xie and Pan 1992)

Item	Eel	Red Nile tilapia	Channel catfish	Freshwater pomfret	Southern catfish	Grass carp
Protein content (%)	16.51	16.94	16.31	18.75	15.28	17.53
Fat content (%)	18.31	0.95	2.99	6.68	1.48	1.8
Essential Amino Acid Index (%)	55.43	70.87	83.79	61.54	63.33	42.76
Flavored amino acid content (%)	19.78	29.34	32.62	25.22	19.57	16.77

content, fat content, essential amino acid index of protein, and flavored amino acid content, which can represent the nutritional value of fish. The nutritional value of eel, red Nile tilapia, channel catfish, freshwater white pomfret, southern catfish, and grass carp was evaluated using gray clustering to compare their nutritional values. The proposed evaluation criteria are shown in Tables 4.7. The specific calculation steps are as follows:

4.6.2.1 Determination of Clustering Factors

According to the items listed in Table 4.7, the cluster analysis uses four factors, namely $x = \{x_1, x_2, x_3, x_4\} = \{$protein content, fat content, Essential Amino Acid Index, flavored amino acid content$\}$. The level of nutrition value was divided into three levels, with low, medium, and high denoted as I, II, and III, respectively.

4.6.2.2 Dimensionless Calculation

For comparison, the influence of the dimensions in Tables 4.7 and 4.8 needs to be eliminated. Therefore, dimensionless calculation is needed. The calculation results are shown in Tables 4.9.

$$M_j = \max x_{ij} \qquad i = 1, 2, \cdots 9$$
$$d_{ij} = x_{ij} / M_j \qquad j = 1, 2, \cdots 4$$

Table 4.9 Nutritional components and standards of the six fish species after nondimensionalization (Xie and Pan 1992)

Protein content (%)	Eel	Red Nile tilapia	Channel Catfish	Freshwater pomfret	Southern catfish	Grass carp	I	II	III
Fat content (%)	0.826	0.847	0.816	0.938	0.764	0.877	0.6	0.75	1
Essential Amino Acid Index(%)	1	0.052	0.163	0.365	0.081	0.098	0.055	0.164	0.546
Flavored amino acid content (%)	0.662	0.846	1	0.735	0.756	0.51	0.239	0.597	0.955
Protein content (%)	0.606	0.899	1	0.773	0.599	0.514	0.307	0.613	0.919

4.6.2.3 Determine the Whitening Functions of Various Types

According to Table 4.9, the threshold λ_{jk} of the whitening function is determined, and the maximum value of f_{ji} is 1.

4.6.2.4 Find the Clustering Weight of Each Item

For example, for the protein content, the low trophic level weight is

$$\eta_{11} = \frac{1/\lambda_{11}}{1/\lambda_{11} + 1/\lambda_{21} + 1/\lambda_{31} + 1/\lambda_{41}} = \frac{1/0.6}{1/0.6 + 1/0.055 + 1/0.239 + 1/0.613} = 0.065$$

The rights of other categories can also be obtained.

4.6.2.5 Calculate the Clustering Coefficient

Let the coefficient of the *i*th sample be σ_{ik}

$$\sigma_{ik} = \sum_{j=1}^{4} f_{jk}(d_{ij})\eta_{jk}$$

For example, for eels, the clustering coefficient of the low trophic level is σ_{11}

$$\sigma_{11} = f_{11}(d_{11})\,\eta_{11} + f_{21}(d_{12})\,\eta_{21} + f_{31}(d_{13})\,\eta_{31} + f_{41}(d_{14})\,\eta_{41} = 0 + 0 + 0.606 \times 0.118 \times (1 - 0.977)) + 0 = 0.002$$

where f_{jk} is obtained by interpolation according to the whitening function. Similarly, the clustering coefficients of eels with respect to other nutritional levels can be obtained.

$$\sigma_{12} = 0.246$$

$$\sigma_{13} = 0.445$$

4.6.2.6 Construct a Clustering Vector and Classify It According to the Maximum Principle

For eel, its clustering vector σ_1

Table 4.10 Clustering coefficients and clustering results of each sample (Xie and Pan 1992)

Grade	Eel	Red Nile tilapia	Channel catfish	Freshwater pomfret	Southern catfish	Grass carp
I	0.002	0.034	0.088	0	0.044	0.08
II	0.246	0.113	0.091	0.254	0.253	0.189
III	0.445	0.374	0.473	0.361	0.079	0.089
Clustering	III	III	III	III	II	II

$$\sigma_1 = (0.002, 0.246, 0.445)$$

$\sigma_{13} = 0.445$ is the maximum, so the eels belong to the high trophic level.

Clustering coefficient of various fishes, the results of σ_{jk} and clustering are shown in Table 4.10.

The clustering results showed that, except for the mesotrophic level of southern catfish and carp, all the other fish were of high trophic level. From high to low: Channel catfish ($\sigma_3 = 0.473$) > Eel ($\sigma_1 = 0.445$) > Red Nile tilapia ($\sigma_2 = 0.374$) > Freshwater Pomfret ($\sigma_4 = 0.361$) > Southern catfish ($\sigma_5 = 0.253$) > silver carp ($\sigma_6 = 0.189$).

The clustering results also showed that the nutritional value of the five famous and high-quality fish was higher than that of the carp, and the nutritional value of the middle-class southern catfish was also higher than that of the carp. Among the five famous fish species, the channel catfish was the highest ($\sigma_3 = 0.473$), and the eel was slightly lower than that ($\sigma_1 = 0.445$). Red Nile tilapia and freshwater pomfret are relatively close to each other ($\sigma_2 = 0.374$, $\sigma_4 = 0.361$).

In addition, according to the survey, the current sales price of aquatic products in China is basically as follows: eel > channel catfish > red Nile tilapia > freshwater white pomfret > southern catfish > grass carp, indicating that the market value of fish mainly depends on the nutritional value. However, the price of eel is higher than that of channel catfish, which may be due to the following factors: Japan has a large demand for eels, and it is impossible to artificially breed eels; different measurement conditions may also cause inconsistency; the size of each type of fish and different feeding methods and different measurement seasons may cause some errors.

In addition to the above components, the nutritional value of fish also includes a variety of vitamins and inorganic salts. Therefore, in future evaluations of the nutritional value of fish, more factors should be selected to meet the actual situation.

4.6.3 Application of Gray Clustering Analysis in the Environmental Assessment of Fishery Waters

Liu and Xiong (2000) published "The Application of Gray Clustering Analysis in the Scientific Management of Reservoir Water Quality." This paper selected TP, TN, and COD as the evaluation factors to evaluate the fishery water quality of Shihe

Table 4.11 Water quality evaluation criteria (Liu and Xiong 2000)

Evaluation factor	I (Very poor trophic level)	II (Poor trophic level)	III (Medium trophic level)	IV (Eutrophic level)	V (Extremely eutrophic level)
TP/μg.L^{-1}	<1	4	23	110	>600
TN/mg.L^{-1}	<0.02	0.06	0.31	1.20	>4.60
COD/mg.L^{-1}	<0.09	0.36	1.80	7.10	>27.10

Table 4.12 Nondimensional water quality evaluation standards (Liu and Xiong 2000)

Evaluation factor	I (Very poor trophic level)	II (Poor trophic level)	III (Medium trophic level)	IV (Eutrophic level)	V (Extremely eutrophic level)
TP/μg.L^{-1}	<0.009	0.036	0.209	1	6
TN/mg.L^{-1}	<0.017	0.05	0.258	1	3.833
COD/mg.L^{-1}	<0.013	0.051	0.254	1	3.817

Reservoir and to grasp the water quality status. The fishery water quality evaluation criteria used in this study are shown in Tables 4.11. The specific calculation steps are as follows:

4.6.3.1 Dimensionless Processing

First, the gray class (evaluation level) and the original whitening number (measured data of each factor) are dimensionlessly processed to facilitate comparison and eliminate the effect of dimension. The dimensionless processing formula of the gray class is $\lambda_{ij} = a_{ij}/a_j$, $i = 1, 2, 3$; $j = 1, 2, 3, 4, 5$. In the formula, λ_{ij} represents the standard value of the whitening number (measured data) after dimensionless treatment; a_i represents the reference standard of the *i*th factor. Here, the standard value of grade IV of each factor is used as the corresponding a_i value. The nondimensional water quality standards are shown in Table 4.12.

4.6.3.2 Calculating the Clustering Weight

Use the formula: $w_{ij} = \lambda_{ij}/\sum_{i=1}^{n} \lambda_{iij}$ The weight of each factor with respect to each level is obtained (Table 4.13).

4.6.3.3 Calculation of the Clustering Coefficient

The clustering coefficient of the *k*th clustering object at the *j*th level is

Table 4.13 The weight of each factor with respect to each level (Liu and Xiong 2000)

Evaluation factor	I (Very poor trophic level)	II (Poor trophic level)	III (Medium trophic level)	IV (Eutrophic level)	V (Extremely eutrophic level)
TP/μg.L^{-1}	0.231	0.263	0.290	0.333	0.440
TN/mg.L^{-1}	0.456	0.365	0.358	0.333	0.281
COD/mg.L^{-1}	0.333	0.372	0.352	0.333	0.280

Table 4.14 Clustering coefficients and results (Liu and Xiong 2000)

I (Very poor trophic level)	II (Poor trophic level)	III (Medium trophic level)	IV (Eutrophic level)	V (Extremely eutrophic level)
0	0	0.770	0.223	0

$$\sigma_{ij} = \sum_{i=1}^{n} f_{ij}(d_{ki})w_{ij}$$

where $k = 1$ and $j = 1, 2, 3$. The method of obtaining $f_{ij}(d_{ki})$is described in Liu and Xiong (2000). The calculation results are shown in Table 4.14.

4.6.3.4 Evaluation Results

The horizontal clustering quantity of Shihe Reservoir is $\sigma = (0, 0, 0.770, 0.223, 0)$, where $\sigma_{13} = 0.770$. it is a medium trophic level.

4.6.4 Application of Gray Clustering in the Classification of Fish Populations

Chen et al. (2002) used the gray variable weight clustering method to preliminarily classify the population structure of *Ommastrephes bartramii* in the Northwest Pacific. In this study, a total of 120 squid samples were randomly collected in the waters west of 165°E in the northwestern Pacific Ocean from October to November 1999. The mantle length (ML), fin length (Q_1), fin width (Q_2), eye diameter (Y), length of the right first arm (WN_1), length of the right second arm (WN_2), length of the right third tentacle (WN_3), length of the right fourth tentacle (WN_4), and length of the right tentacle spike (SL) were determined. The cluster analysis method of gray variable weights was used to study the population structure of *Ommastrephes bartramii* in the northwestern Pacific Ocean.

Table 4.15 Clustering results of each sample (Chen et al. 2002)

Category	Sample	Total
Class I	1, 2, 3, 4, 5, 6, 7, 8, 9, 10, 11, 12, 13, 14, 15, 16, 17, 18, 19, 20, 21, 22, 23, 25, 26, 29, 30, 31, 32, 33, 35, 38, 41, 42, 45, 46, 47, 48, 49, 50, 52, 53, 54, 55, 56, 57, 58, 59, 60, 61, 62, 63, 64, 66, 68, 69, 70, 71, 73, 74, 75, 76, 77, 78, 79, 80, 81, 82, 83, 84, 85, 86, 88, 89, 91, 92, 93, 94, 95, 96, 97, 98, 99, 100, 101, 102, 103, 104, 105, 106, 107, 108, 110, 112, 115, 116, 117, 118, 120	99
Class II	24, 27, 28, 34, 36, 37, 39, 40, 43, 44, 51, 65, 67, 72, 87, 90, 109, 111, 113, 114, 119	21

According to the 9 collected morphological characteristic indexes, the ratios of other metrics to mantle length were obtained, i.e., Q_1/ML, Q_2/ML, Y/ML, W_1/ML, W_2/ML, W_3/ML, W_4/ML, and S/ML, and a matrix was constructed with 8 indicators as columns and 120 samples as rows.

$$d_{ij} = \begin{vmatrix} d_{1,1}(0.425) & d_{1,2}(0.766) & \cdots & d_{1,8}(0.492) \\ d_{2,1}(0.428) & d_{2,2}(0.736) & \cdots & d_{2,8}(0.553) \\ \vdots & \vdots & \vdots & \vdots \\ d_{120,1}(0.444) & d_{120,2}(0.791) & \cdots & d_{120,8}(0.528) \end{vmatrix}$$

Three indicators of high, medium, and low levels were used to set the gray-like whitening function. The clustering coefficient was obtained, and the clustering coefficient vector was constructed. The clustering results are shown in Table 4.15. The results of the study are as follows:

1. The gray variable weight clustering method was used to analyze the population structure of *Ommastrephes bartramii* in the northwestern Pacific Ocean (Table 4.15).
2. The calculation showed that there was a significant difference in the average values of the eight morphological characteristics of the two populations (Table 4.16). The variation in the coefficient of variation was between 2.54% and 9.15%. The most significant indicator was the ratio of the length of each arm to the mantle length, while the coefficient of difference in the ratio of fin length and fin width to mantle length was relatively small.
3. The conclusions reached are basically the same as those of previous studies. The use of morphological analysis methods requires quantitative biological measurement, and sampling and measurement are simple and easy to implement. At the same time, using the gray clustering analysis method, its mathematical processing method is relatively simple and scientific. However, the morphological feature value is easily affected by environmental factors, which reduces the stability of the feature itself and affects the credibility of the identification results (Chen et al. 2002).

Table 4.16 Average morphological characteristics of the two groups (Chen et al. 2002)

Class I	Indicator	S_{Q1}/S_{ML}	S_{Q2}/S_{ML}	$S\ Y/S_{ML}$	S_{W1}/S_{ML}
	Mean value	0.4372	0.7717	0.0785	0.4929
Class II	Indicator	S_{Q1}/S_{ML}	S_{Q2}/S_{ML}	$S\ Y/S_{ML}$	S_{W1}/S_{ML}
	Mean value	0.4223	0.7521	0.0723	0.4545
Coefficient of difference		3.41%	2.54%	7.90%	7.80%
Class I	Indicator	S_{W2}/S_{ML}	S_{W3}/S_{ML}	S_{W4}/S_{ML}	S_{SL}/S_{ML}
	Mean value	0.5847	0.6256	0.5846	0.6016
Class II	Indicator	S_{W2}/S_{ML}	S_{W3}/S_{ML}	S_{W4}/S_{ML}	$S\ SL/S_{ML}$
	Mean value	0.5449	0.5801	0.5311	0.5561
Coefficient of difference		3.41%	6.81%	7.27%	9.15%

References

Chen XJ (2003) Application of Gray system theory in fishery science. China Agricultural Press. (In Chinese)

Chen XJ (2023) Application of Gray system theory in fishery science. China Agricultural Press. (In Chinese)

Chen XJ, Zhang XG (2001) Preliminary discussion on regional types of Chinese fisheries economy in coastal provinces. J Shanghai Fish Univ 10(2):183–186. (In Chinese)

Chen XJ, Tian SQ, Ye XC (2002) Study on population structure of flying squid in Northwestern Pacific based on gray system theory. J Shanghai Fish Univ 11(4):335–341. (In Chinese)

Liu DQ, Xiong BX (2000) Application of gray cluster analysis in scientific management of reservoir water quality. J Hydroecol 20(3):41–47. (In Chinese)

Liu SF, Yang YJ, Wu LF et al (2014) Gray system theory and its application. Science Press, Beijing. (In Chinese)

Xie J, Pan HJ (1992) Gray system comprehensive evaluation of nutritional value of several famous fishes. Zhujiang Aquatic (19):67–72. (In Chinese)

Chapter 5
Basic Principles of Gray Dynamic Modeling

Xinjun Chen

Abstract Gray dynamic modeling is one of the important components of gray system theory. Gray system is a kind of system which contains not only known information, but also unknown information or non-known information. One of the important contents of studying the gray system is how to abstract and build a model from an ambiguous system with insufficient overall information, which can make the factors of the gray system from unclear to clear, from little knowledge to more knowledge provides the basis for research. In the theory of gray system, a model that describes the continuous changing process of things in a gray system is established by generating and transforming the original data sequence that represents the behavior characteristics of the gray system, called the gray model. Gray models are the long-term fuzzy description of the development of things by establishing the gray differential prediction model with a little and incomplete information. The basic idea of its modeling is to use the original data to form the original sequence (0) and to generate the sequence (1) by the method of accumulation. The generated sequence (1) can weaken the randomness of the original data and make it present more obvious characteristic law. The differential equation model is creat by use of the generated sequence (1), i.e. GM model. The GM (1,1) model represents a first-order, one-variable differential equation model. The gray model usually adopts the "five-step modeling" method, which includes system qualitative analysis, factor analysis, preliminary quantification, dynamic quantification and optimization, it has the advantages of not needing a large number of samples and not requiring regular distribution of samples. Generally speaking, social system, economic system, ecological system, and fishery system are all gray systems. For example, there are many factors that lead to the decline of fishing yield, but not many are known, so the gray forecasting method can be used for the prediction of the gray system of fishing yield. In this chapter, we will mainly introduce the technical route of gray modeling,

---

X. Chen (✉)
College of Marine Sciences, Shanghai Ocean University, Lingang New City, Shanghai, China
e-mail: xjchen@shou.edu.cn

X. Chen (ed.), *Application of Gray System Theory in Fishery Science*,
https://doi.org/10.1007/978-981-99-0635-2_5

including the principle of gray dynamic modeling and the modeling process and steps of common GM (1, 1) and GM (1, *n*) models.

Keywords Gray dynamic modeling · GM (1,1) · GM (1, *N*) · Gray system

5.1 Technical Route of Gray Modeling

5.1.1 Principle of Gray Dynamic Modeling

Gray predictive modeling is based on the concept of gray modules. Gray system theory believes that all random quantities are gray quantities and gray processes that vary within a certain range and within a certain period of time. The processing of the gray quantity is not to seek its statistical law and probability distribution but to find the law from the irregular raw data, that is, to process the data in a certain way to make it into more regular time series data. Then, the model is established. This is because in an objective system, no matter how complex it is, there is always correlation, overall function, and order within the system. Therefore, as the data that characterize the system's behavior, it always contains a certain pattern. The sequence data generated by processing in a certain way are called "modules." Its geometric meaning is the general term formed by the continuous curve and its bottom (i.e., the abscissa) of the generated sequence data on the two-dimensional plane of time and data. We call the module composed of known data columns the white module, and the module that is extrapolated from the white module to the future, that is, the module composed of the predicted value, is called the gray module.

Under normal circumstances, for a given raw data column

$$X_{(0)} = \left\{ x_{(0)}^{(1)}, x_{(0)}^{(2)}, x_{(0)}^{(3)}, \ldots, x_{(0)}^{(N)} \right\}$$

It cannot be directly used for modeling because these data are mostly random and irregular. If the original data column is generated by one accumulation, a new data column can be obtained.

$$X_{(1)} = \left\{ x_{(1)}^{(1)}, x_{(1)}^{(2)}, x_{(1)}^{(3)}, \cdots, x_{(1)}^{(N)} \right\}$$

where $x_{(1)}^{(i)} = \sum_{k=1}^{i} x_{(0)}^{(i)}$

The newly generated data sequence is a monotonic growth curve, which obviously enhances the regularity of the original number sequence, while the randomness is weakened. For a nonnegative data series, the greater the number of accumulations, the more obvious the weakening of randomness, and the stronger the regularity, so it is easier to approximate the exponential function. The processed data weaken the randomness of the original data series, thereby finding the regularity of its variation and providing intermediate information for the establishment of dynamic models.

For biology, social economics, and environmental ecosystems, the application of differential equations facilitates the description of the dynamic characteristics of their internal physical or chemical processes. Gray system theory can be used to establish a differential equation model because gray system theory treats a random quantity as a gray quantity that changes within a certain range and treats a stochastic process as a gray process that changes in a certain range and a certain time zone. Second, gray system theory generates irregular raw data and turns it into a more regularly generated sequence of numbers for modeling.

Therefore, gray GM modeling is actually a generated data model, while the original data model is generally used for modeling. In addition, gray system theory adjusts, corrects, and improves the accuracy of the model through different generations of gray numbers, different selections of data, and the supplementation of different levels of residual models.

5.1.2 *Common GM* (n, h) *Model*

The GM (n, h) model refers to a differential equation with n-order h variables. Different GM models with n and h have different meanings and uses. The GM model can be roughly classified into the following two categories.

5.1.2.1 GM (n, 1) Model

The GM (n, 1) model is commonly used, that is, the GM model with only one variable. The requirement for the data column is the time series of the "comprehensive effect." Because the larger n is, the more complex the calculation, the accuracy may not be higher. Therefore, in general, the value of n is below the third order. The most commonly used $n = 1$-order model is simple in calculation and has wide applicability.

The differential equation of the GM (1, 1) model is $\frac{dx^{(1)}}{dt} + ax^{(1)} = u$.

Coefficient vector $\widehat{a} = [a, \mu]^T$.

The corresponding time function is $\widehat{x}^{(1)}(t+1) = \left(x^{(0)}(1) - \frac{u}{a}\right)e^{-at} + \frac{u}{a}$.

After derivation and reduction, we can obtain $\widehat{x}^{(0)}(t+1) = -a\left(x^{(0)}(1) - \frac{u}{a}\right)e^{-at}$.

The above two equations are the basic calculation formulas for the gray prediction of the GM (1, 1) model.

GM (2, 1) is a second-order model with two characteristic roots. Its dynamic process can reflect different situations; that is, it may be monotonic, nonmonotonic, or oscillating.

The differential equation of the GM (2, 1) model is

$$\frac{d^2x^{(1)}}{dt^2} + a_1\frac{dx^{(1)}}{dt} + a_2x^{(1)} = u$$

Coefficient vector $\widehat{a} = (a_1, a_2, u)^T$The time response function is

$$x^{(1)}(t) = C_1 e^{\lambda_1 t} + C_2 e^{\lambda_2 t} + \frac{u}{a^2}$$

where λ_1,λ_2 are the two characteristic roots, and the main dynamic characteristics of the system can be analyzed according to the following different situations (Deng 1990; Liu et al. 2014).

1. If $\lambda_1 = \lambda_2$, then the dynamic process is monotonic.
2. If $\lambda_1 \neq \lambda_2$ is a real number, the dynamic process may be nonmonotonic.
3. If $\lambda_1.\lambda_2$ is the conjugate complex root, then the dynamic process is cyclically oscillating.

5.1.2.2 GM (1, *h*) Model

The GM (1, 1) and GM (2, 1) described above are generally used for prediction. As a state analysis model, the GM (1, *h*) model is commonly used, which can reflect the effect of $h - 1$ variables on the first derivative of the dependent variable. Since $h > 1$, it is called the first-order linear dynamic model of *h* sequences. The modeling steps are as follows (Deng 1990; Chen 2003; Liu et al. 2014):

Let *h* variables $X_1, X_2, \ldots, X_h$ form the original sequence of numbers $x_i^{(0)} = \{x_i^{(0)}$ (1), $x_i^{(0)}$ (2), …, $x_i^{(0)}$ $(N)\}$ $(i = 1, 2, \ldots, h)$. A new sequence of numbers is obtained by accumulating $X_i^{(0)}$ once:

$$X_i^{(1)} = \left\{x_i^{(1)}(1), x_i^{(1)}(2), \ldots, x_i^{(1)}(n)\right\} \; (I = 1, 2, \ldots, h)$$

Establishment of the differential equation:

$$\frac{dx_1^{(1)}}{dt} + ax_1^{(1)} = b_1 x_2^{(1)} + b_2 x_3^{(1)} + \cdots + b_{h-1} x_h^{(1)}$$

Coefficient vector $\hat{a} = (b_1, b_2, \ldots, b_h - 1)^T$ is solved using the least squares method, i.e.,

$$\hat{a} = \left(\boldsymbol{B}^T B\right)^{-1} B^T Y_N$$

where $\boldsymbol{B}$ is the cumulative matrix and Y_N is the vector of the constant terms:

$$\boldsymbol{B} = \begin{bmatrix} -\frac{1}{2}\left(x^{(1)}(1) + x^{(1)}(2)\right) & x_2^{(1)}(2) & \cdots & x_h^{(1)}(2) \\ -\frac{1}{2}\left(x^{(1)}(2) + x^{(1)}(3)\right) & x_2^{(1)}(3) & \cdots & x_h^{(1)}(3) \\ \cdots & \cdots & \cdots & \cdots \\ -\frac{1}{2}\left(x^{(1)}(n-1) + x^{(1)}(n)\right) & x_2^{(1)}(n) & \cdots & x_h^{(1)}(n) \end{bmatrix}$$

$$Y_N = \left[x_1{}^{(0)}(2), x_1{}^{(0)}(3), \ldots, x_1{}^{(0)}(n)\right]^T$$

Then, the solution of the differential equation can be obtained:

$$x_1^{(1)}(t+!) = \left(x_1^{(0)}(1) - \sum_{i=2}^{h} \frac{b_{i-1}}{a} x_i^{(1)}(t+!)\right) e^{-at} + \sum_{i=2}^{h} \frac{b_{i-1}}{a} x_i^{(1)}(t+!)$$

5.2 GM (1, 1) Model

The GM (1, 1) model is actually the prediction of the gray series, which carries out the quantitative prediction of the time series data, such as population prediction, labor prediction, output prediction, output value prediction, and various trend predictions. Its future development is predicted. This type of prediction is not only widely used but also has universal significance. To this end, we provide a more detailed introduction.

5.2.1 The Basic Steps of Establishing the GM (1, 1) Model

Step 1: The data sequence $X^{(0)} = \{x^{(0)}(1), x^{(0)}(2), \ldots, x^{(0)}(N)\}$ is accumulated and generated once to obtain

$$X^{(1)} = \left\{x^{(1)}(1), x^{(1)}(2), \ldots, x^{(1)}(N)\right\}$$

where $x^{(1)}(t) = \sum_{k=1}^{t} x^{(0)}(k)$

Step 2: Construct the accumulation matrix B and the constant term vector Y_N, i.e.,

$$\boldsymbol{B} = \begin{bmatrix} & 1 \\ -\frac{1}{2}\left(x^{(1)}(1) + x^{(1)}(2)\right) & 1 \\ -\frac{1}{2}\left(x^{()}(2) + x^{(1)}(3)\right) & \\ \vdots & \vdots \\ -\frac{1}{2}\left(x^{(1)}(N-) + x^{(1)}(N)\right) & \\ & 1 \end{bmatrix}$$

$$Y_N = \left[x_1{}^{(0)}(2), x_1{}^{(0)}(3), \ldots, x_1{}^{(0)}(N)\right]^T$$

Step 3: Solve the gray parameters using the least squares method $\widehat{a}$

$$\widehat{a} = \begin{bmatrix} a \\ u \end{bmatrix} = \left(B^T B\right)^{-1} B^T Y_N$$

Step 4: Substitute the gray parameters into the time function:

$$\widehat{x}^{(1)}(t+1) = \left(x^{(0)}(1) - \frac{u}{a}\right)e^{-at} + \frac{u}{a}$$

Step 5: Taking the derivative of $\widehat{X}^{(1)}$, we get the following

$$\widehat{x}^{(0)}(t+1) = -a\left(x^{(0)}(1) - \frac{u}{a}\right)e^{-at}$$

or

$$\widehat{x}^{(0)}(t+1) = \widehat{x}^{(1)}(t+1) - \widehat{x}^{(1)}(t)$$

Step 6: Calculating the difference $\varepsilon^{(0)}(t)$ between $x^{(0)}(t)$ and $\widehat{x}^{(0)}(t)$, and the relative error $e(t)$

$$\varepsilon^{(0)}(t) = x^{(0)}(t) - \widehat{x}^{(0)}(t)$$
$$e(t) = \varepsilon^{(0)}(t)/x^{(0)}(t)$$

Step 7: Testing the model accuracy and application of the model for prediction.

To analyze the reliability of the model, the accuracy of the model must be tested. Currently, the most common diagnostic method is to perform a posterior error test on the model. That is, the deviation s_1 of the observation data is first calculated:

$$s_1^2 = \sum_{t=1}^{m} \left(x^{(0)}(t) - \overline{x}^{(0)}(t)\right)^2$$

Residual deviation s_2:

$$s_2^2 = \frac{1}{m-1} \sum_{t=1}^{m-1} \left(q^{(0)}(t) - \overline{q}^{(0)}(t)\right)^2$$

Then, calculate the posterior ratio: $c=\frac{s_1}{s_2}$ and the probability of small error: $p=\left\{\left|q^{(0)}(t)-\overline{q}^{(0)}\right|<0.6745s_1\right\}$

The model was diagnosed based on the posterior ratio c and the small error probability p. When $p > 0.95$ and $c < 0.35$, the model can be considered reliable and can be used for prediction. At this time, the system behavior can be predicted based on the model.

The above seven steps are the entire modeling and prediction analysis process. When the residual of the established model is large and the accuracy is not ideal, to improve the accuracy, residual GM (1, 1) model modeling analysis should be performed to correct the prediction model.

5.2.2 *Different Forms of the GM (1, 1) Model*

Based on the results of Professor Deng's research on the GM (1, 1) model, a variety of different forms of the GM (1, 1) model are proposed, and the main ones are listed as follows: (Deng 1990; Chen 2003; Liu et al. 2014):

1. $\begin{cases} x^{(0)}(2)=\beta-\alpha x^{(0)}(1) \\ x^{(0)}(k)=(1-\alpha)x^{(0)}(k-1); k=3,4,\cdots n \end{cases}$
2. $\begin{cases} x^{(0)}(2)=\beta-\alpha x^{(0)}(1) \\ x^{(0)}(k)=\beta-\alpha x^{(0)}(k-1); k=3,4,\cdots n \end{cases}$
3. $\begin{cases} x^{(0)}(2)=\beta-\alpha x^{(0)}(1) \\ x^{(0)}(k)=\dfrac{x^{(1)}(k-1)-0.5b}{x^{(1)}(k-2)+0.5b}x^{(0)}(k-1); k=3,4,\cdots n \end{cases}$
4. $x^{(0)}(k)=\left(\frac{1-0.5a}{1+0.5a}\right)^{k-2}\left(\frac{b-ax^{(0)}(1)}{1+0.5a}\right); k=2,3,\ldots,n$
5. $x^{(0)}(k)=\frac{1}{(1-\alpha)^3}x^{(0)}(3)e^{k\ln(1-\alpha)}, k>3$
6. $x^{(0)}(k)=(\beta-\alpha x^{(0)}(1))e^{-\alpha(k-2)}$
7. $\widehat{x}^{(0)}(k)=(1-e^{a})\left(x^{(0)}(1)-\frac{b}{a}\right)e^{-a(k-1)}$
8. $\widehat{x}^{(0)}(k)=(-a)\left(x^{(0)}(1)-\frac{b}{a}\right)e^{-a(k-1)}$

5.2.3 *Interpretation and Analysis of the Development Coefficient-a Value*

The parameter *-a* in the GM (1, 1) model is usually referred to as the development coefficient, and the parameter *b* is referred to as the gray effect (Deng 1990; Liu et al. 2014).

-a reflects $\widehat{x}^{(0)}$ and $\widehat{x}^{(1)}$ the development trend. In general, the system action quantity should be exogenous or predetermined, while GM (1, 1) is a single-sequence modeling, and only the behavior sequence of the system (called the output

sequence) is used, and there is no external action sequence (or called the input sequence). In GM (1, 1), the gray effect is the data mined from the background value, which reflects the relationship of data changes, and its exact connotation is gray. The existence of gray action quantity is a concrete manifestation of connotative extension. Its existence is the watershed that distinguishes gray modeling from general input–output modeling and is also an important sign to distinguish the gray system view from the gray box view.

In addition, Liu et al. (2014) conducted an in-depth study on the development coefficient *-a* in the GM (1, 1) model by analyzing the simulation error and prediction error of GM (1, 1), and the magnitude and value of the development coefficient *-a* were compared. The prediction accuracy of the system and its possible application are discussed, and the following conclusions are drawn:

1. When $-a \le 0.3$, GM (1, 1) can be used for medium- and long-term prediction;
2. When $0.3 < -a \le 0.5$, it can be used for short-term prediction, and medium- and long-term prediction can be used with caution;
3. When $0.5 < -a \le 0.8$, caution should be taken when using GM (1,1) for short-term prediction;
4. When $0.8 < -a \le 1$, the residual correction GM (1, 1) model should be used;
5. When $-a > 1$, the GM (1, 1) model should not be used.

5.3 GM (1, *n*) Model of Gray Sequence

The GM (1, 1) model and the GM (2, 1) model are both single-sequence linear dynamic models, while the GM (1, *n*) model is a multivariate (multivariate) first-order linear dynamic model. It is mainly used for system dynamic analysis (Deng 1990; Chen 2003; Liu et al. 2014).

For the original data with *n* sequences and a sequence length of *m*, we can use the following data matrix to describe:

$$X_N^{(0)} = \begin{bmatrix} x_1^{(0)}(1) & x_2^{(0)}(1) & \cdots & x_n^{(0)}(1) \\ x_1^{(0)}(2) & x_2^{(0)}(2) & \cdots & x_n^{(0)}(2) \\ \cdots & \cdots & \cdots & \cdots \\ x_1^{(0)}(m) & x_2^{(0)}(m) & \cdots & x_n^{(0)}(m) \end{bmatrix}$$

Step 1: Calculate the data matrix generated by one accumulation

$$X_N^{(1)} = \begin{bmatrix} \sum_{i=1}^{2} x_1^{(0)}(i) & \sum_{i=1}^{2} x_2^{(0)}(i) & \cdots & \sum_{i=1}^{2} x_N^{(0)}(i) \\ \sum_{i=1}^{3} x_1^{(0)}(i)) & \sum_{i=1}^{3} x_2^{(0)}(i) & \cdots & \sum_{i=1}^{3} x_N^{(0)}(i) \\ \cdots & \cdots & \cdots & \cdots \\ \sum_{i=1}^{M} x_1^{(0)}(i) & \sum_{i=1}^{M} x_2^{(0)}(i) & \cdots & \sum_{i=1}^{M} x_N^{(0)}(i) \end{bmatrix}$$

Step 2: Construct the matrix ***B***, *Y* and the adjacent mean value to generate the sequence Z_i

$$B = \begin{bmatrix} -\frac{1}{2}\left(x_1^{(1)}(2) + x_1^{(1)}(1)\right) & x_2^{(1)}(2) & \cdots & x_n^{(1)}(2) \\ -\frac{1}{2}\left(x_1^{(1)}(3) + x_1^{(1)}(2)\right) & x_2^{(1)}(3) & \cdots & x_n^{(1)}(3) \\ \cdots & \cdots & \cdots & \cdots \\ -\frac{1}{2}\left(x_1^{(1)}(m) + x_1^{(1)}(m-1)\right) & x_2^{(1)}(m) & \cdots & x_n^{(1)}(m) \end{bmatrix}$$

$$Y = \left[x_1{}^{(0)}(2), x_1{}^{(0)}(3) \ldots, x_1{}^{(0)}(m)\right]^T$$

$$Z = \begin{pmatrix} -\frac{1}{2}\left(x_i^{(1)}(2) + x_i^{(1)}(1)\right) \\ -\frac{1}{2}\left(x_i^{(1)}(3) + x_i^{(1)}(2)\right) \\ \cdots \\ -\frac{1}{2}\left(x_i^{(1)}(m) + x_i^{(1)}(m-1)\right) \end{pmatrix}$$

Step 3: Solve the gray parameters using the least squares method $\widehat{a}$

$$\widehat{a} = \begin{bmatrix} a \\ b_1 \\ \vdots \\ b_{n-1} \end{bmatrix} = \left(B^T B\right)^{-1} B^T Y$$

Step 4: Establish the gray differential equation GM (1, *n*)

$$x_1^0(t) + a z_1^{(1)}(t) = \sum_{i=2}^{n} b_i x_i^{(1)}(t)$$

where -*a* is the system development coefficient; $b_i x_i^{(1)}(t)$ is the driving term; b_i is the driving coefficient

Step 5: Substitute the gray parameters into the time function

$$\hat{x}^{(1)}(t+1) = \left(x^{(0)}(1) - \sum_{i=2}^{n} \frac{b_i - 1}{a} x_i^{(1)}(t+1) \right) e^{-at} + \sum_{i=2}^{n} \frac{b_i - 1}{a} x_i^{(1)}(t+1)$$

Step 6: Substitute the gray parameters into the time function to obtain the calculated values of the generated data series $\hat{x}^{(1)}(t)$. Then, taking the derivative of $\hat{x}^{(1)}$, we obtain $\hat{x}^{(0)}(t)$. The difference $\varepsilon^{(0)}(t)$ between $x^{(0)}(t)$ and $\hat{x}^{(0)}(t)$, and the relative error $e(t)$ are calculated.

Step 7: While establishing the model, the system also performs Laplace transformation on the parameters of the GM $(1, n)$ model and gives the dynamic link transfer function $w_i\ (s)$ of the ith influencing factor to its action object under the zero initial condition.

$$w_i(s) = \frac{x_1^{(1)}(s)}{x_i^{(1)}(s)} = \frac{b_i/a}{1 + 1/as} \qquad (i = 2, 3, \ldots, n)$$

References

Chen XJ (2003) Application of gray system theory in fishery science. China Agricultural Press. (In Chinese)

Deng JL (1990) A course in gray system theory. Central China Science and Technology University Press

Liu SF, Yang YJ, Wu LF et al (2014) Gray system theory and its application. Science Press, Beijing. (In Chinese)

Chapter 6
Gray Prediction

Xinjun Chen, Minyang Xie, and Jintao Wang

Abstract Prediction is a kind of activity that makes use of the knowledge and means that people have already mastered to predict and judge the future development of things. Specifically, people use various qualitative and quantitative analysis methods according to the objective process and certain laws of the development and change of things in the past, and according to the state of movement and change of things, a scientific projection of the possible future trends and possible levels of things. As a kind of human cognitive activity, prediction has existed in human social practice for a long time and has been developing with the development of productivity and Relations of production. Therefore, the forecast is actually by means of the past to predict and understand the future trend of development. Usually, prediction can be divided into qualitative prediction and quantitative prediction, quantitative prediction is through the analysis of data for the prediction, often need to establish a prediction model. Gray prediction is to discover and grasp the law of system development by processing the original data and establishing the gray model and to make a scientific quantitative prediction of the future state of the system. The gray prediction model does not use the original data sequence, but the generated data sequence. Its core system is gray model, that is, the method of getting approximate exponential law from the original data by accumulative generation (or other processing generation) and then modeling. The advantage of gray prediction is that it can solve the problems of less historical data, integrity of sequence and low reliability without enough sample space of data, it can generate irregular original data to get regular generating sequence. The disadvantage is that it applies only to short-and medium-term forecasts and only to those approximating exponential growth. There should be sufficient quantitative analysis as to what model should be chosen for a particular problem. The choice of models, however, is not inflexible. A model must go through a variety of tests to determine whether it is reasonable, only through the test of the model can be used as a prediction. In this chapter, we will mainly introduce the test method of the gray prediction model, the sequence

X. Chen (✉) · M. Xie · J. Wang
College of Marine Sciences, Shanghai Ocean University, Lingang New City, Shanghai, China
e-mail: xjchen@shou.edu.cn; jtwang@st.shou.edu.cn

X. Chen (ed.), *Application of Gray System Theory in Fishery Science*,
https://doi.org/10.1007/978-981-99-0635-2_6

prediction, the gray catastrophe prediction, and the application example of the gray prediction in fishery science.

Keywords Gray prediction · Gray catastrophe prediction · Fishery science · Gray prediction test

6.1 Test Method of the Gray Prediction Model

6.1.1 Absolute Correlation Test Method

The original sequence is now available

$$X^{(0)} = \left(x^{(0)}(1), x^{(0)}(2), \ldots, x^{(0)}(n)\right)$$

The corresponding relative error sequence is

$$\Delta = \left(\left|\frac{\varepsilon(1)}{x^{(0)}(1)}\right|, \left|\frac{\varepsilon(2)}{x^{(0)}(2)}\right|, \ldots, \left|\frac{\varepsilon(n)}{x^{(0)}(n)}\right|\right) = \{\Delta_k\}_1^n$$

In the above formula, $X^{(0)}$ is the original sequence, $\widehat{X}^{(0)}$ is the corresponding simulation sequence, and ε is the absolute correlation of the corresponding simulation sequence $X^{(0)}$ with $\widehat{X}^{(0)}$. If for a given $\varepsilon_0 > 0$, there is $\varepsilon > \varepsilon_0$, then the model is said to be a qualified model.

6.1.2 Mean Square Error Ratio and Small Error Probability Test Method

Let $X^{(0)}$ is the original sequence, $\widehat{X}^{(0)}$ is the corresponding simulation sequence, and $\varepsilon^{(0)}$ is the residual sequence, the mean and variance of $X^{(0)}$ are

$$\overline{x} = \frac{1}{n}\sum_{k=1}^{n} x^{(0)}(k)$$

$$S_1^2 = \frac{1}{n}\sum_{k=1}^{n}\left(x^{(0)}(k) - \overline{x}\right)^2$$

The mean and variance of the residuals are

Table 6.1 Accuracy test grade reference table (Liu et al. 1999)

Grade	Relative error α	Gray correlation degree ε_0	Mean square error ratio C_0	Probability of small error p_0
Level 1	0.01	0.90	0.35	0.95
Level 2	0.05	0.80	0.5	0.80
Level 3	0.10	0.70	0.65	0.70
Level 4	0.20	0.60	0.80	0.60

$$\overline{\varepsilon}=\frac{1}{n}\sum_{k=1}^{n}\varepsilon(k)$$

$$S_2^2=\frac{1}{n}\sum_{k=1}^{n}(\varepsilon(k)-\overline{\varepsilon})^2$$

1. If in which $C=\frac{S_2}{S_1}$ is called the variance ratio. For a given $C_0>0$, when $C<C_0$, the model is called a qualified model with a mean square error ratio.
2. If $p=P(|\varepsilon(k)-\overline{\varepsilon}|<0.6745S_1)$ is called the probability of a small error. For a given $p_0>0$, when $p>p_0$, the model is called a small error probability qualified model.

Through the above analysis, three methods for testing the model are given. These three methods are all judged by the accuracy of the model by examining the residuals, in which the smaller the average relative error $\overline{\Delta}$ is, the better, and the larger the gray correlation degree ε is. Meanwhile, the smaller the mean square error ratio C is, the better, and the larger the probability of a small error p is, the better. Given a set of values of α, ε_0, C_0, and p_0, a level of simulation accuracy of the test model is determined. The commonly used accuracy grades are shown in Table 6.1. Under normal circumstances, the most commonly used index is the relative error test.

6.2 Sequence Prediction

Sequence prediction predicts the future behavior of system variables. The commonly used sequence prediction model is the GM (1, 1) model. According to the actual situation, other gray models can also be considered. On the basis of qualitative analysis, an appropriate sequence operator is defined, and then a GM (1, 1) model is established. After passing the accuracy test, it can be used for prediction. The entire modeling method can be referred to as the GM (1, 1) model in Chap. 5.

6.3 Prediction of Gray Catastrophe

6.3.1 Gray Catastrophe Prediction

Gray catastrophe prediction is essentially the prediction of outliers. What kind of value is considered an outlier is often determined by the supervisor's experience and historical values. The task of gray catastrophe is to give the time when the next or several outliers appear so that people can prepare in advance and take preventive countermeasures.

Now let X be the original sequence

$$X_\xi = (x[q(1)], x[q(2)], \ldots, x[q(m)])$$

is a catastrophic sequence, then it is called a catastrophic sequence.

$$Q^{(0)} = (q(1), q(2), \ldots, q(m))$$

is the catastrophe date sequence.

Catastrophe prediction searches for the regularity of the catastrophe date series and predicts the dates of several future catastrophes. The catastrophe prediction of the gray system is achieved by establishing a GM (1, 1) model for the catastrophe date series.

Let $Q^{(0)} = (q(1), q(2), \ldots, q(m))$ be the catastrophe date sequence, and its one-time cumulative sequence is

$$Q^{(1)} = (q(1), q(2), \ldots, q(m))$$

The sequence of generating the mean value of $Q^{(1)}$ is called $Z^{(1)}$; then, $q(k) + az^{(1)}(k) = b$ is called the catastrophic GM (1, 1) model.

Now let $X = (x(1), x(2), \ldots, x(n))$ be the original sequence, and let n be the date. Given a certain outlier ξ, the corresponding catastrophe date series

$$Q^{(0)} = (q(1), q(2), \ldots, q(m)),$$

where $q(m)$ ($\leq n$) is the date of the most recent catastrophe; then, $\widehat{q}(m+1)$ is the prediction date of the next catastrophe. For any $k > 0$, we call it $\widehat{q}(m+k)$ the predicted date of the kth catastrophe in the future.

6.3.2 Prediction of Gray Seasonal Catastrophe

Let $\Omega = [a, b]$ be the total time zone. If $\omega_i = [a_i, b_i] \subset [a, b]$ $(i = 1, 2, \ldots, s)$ satisfies $\Omega = \bigcup_{i=1}^{s} \omega_i$; $\Omega_i \cap \omega_j = \varnothing$, and any $j \neq i$, then ω $(i = 1, 2, \ldots, s)$ is called the season in Ω, which is also called the time period or time-sharing zone.

Let $\omega_i \subset \Omega$ be a season, and let the original series

$$X = (x(1), x(2), \ldots, x(n))$$

A given outlier ξ is called the corresponding catastrophe sequence.

$$X_\xi = (x[q(1)], x[q(2)], \ldots, x[q(m)])$$

Correspondingly, we call

$$Q^{(0)} = (q(1), q(2), \ldots, q(m))$$

as the seasonal catastrophe date sequence.

Seasonal catastrophe prediction can be carried out according to the following steps:

Step 1: Given the original sequence $X = (x(1), x(2), \ldots, x(n))$;

Step 2: Study the variation range of the original series data and determine the seasonal $\omega_i = [a_i, b_i]$;

Step 3: Let $y(k) = x(k)\text{-}a_i$, and convert the original sequence to $Y = (y(1), y(2), \ldots, y(n))$ to improve the data resolution;

Step 4: Given the outlier ξ, find the seasonal catastrophe sequence

$$Y_\xi = (y[q(1), q(2), \ldots, q(m)])$$

and seasonal catastrophe date series

$$Q^{(0)} = (q(1), q(2), \ldots, q(m));$$

Step 5: Establish a catastrophe GM (1, 1) model, $q(k) + az^{(1)}(k) = b$;

Step 6: Test the simulation accuracy and make predictions.

6.4 Application of Gray Prediction in Fishery Science

Gray series prediction has been widely used in fishery science. At present, the application of gray prediction in fisheries is mainly in the following aspects: fishery yield, fishery population, fisheries forecasting (including resource abundance and catch), fishery disease, etc.

6.4.1 Gray Prediction of Fishery Yield

6.4.1.1 Gray Prediction of Marine Catch in the Indian Ocean

Lu et al. (2022) published "Construction of a prediction model for marine catches in the Indian Ocean based on the gray system theory model." Based on the catch data of the Indian Ocean from 2000 to 2016, the gray system theory method was used to analyze the main catch categories that affected the total catch, and a variety of GM models were established and compared. The data from 2018 were verified, and the obtained optimal GM model was used to predict the total catch in the Indian Ocean from 2019 to 2025. The results showed that the main categories that affected the total catch in the Indian Ocean were demersal fish, crustaceans, pelagic fish, other marine fishes, and cephalopods. The models were GM (1, 5) and GM (1, 6), with average relative errors of 1.83% and 1.90%, respectively, and the gray correlation degree was above 0.9. The average relative errors of the forecasts in 2017 and 2018 were 3.78% and 3.42%, respectively. The predicted total catches in the Indian Ocean in 2019–2020 and 2021–2025 are 11.86–12.9 million tons and 12.27–13.24 million tons, respectively. The main increase in the catch may come from pelagic fish and cephalopods and demersal fish. The study suggests that the growth rate of the total catch in the Indian Ocean during the Fourteenth Five-Year Plan period is limited, and the total increase is within 800,000 t, which is basically at the stage of full development. It is recommended that the scale of fishery development be strictly controlled in the future to ensure the sustainable development and sustainable use of fishery resources.

The catch data of marine fishing in the Indian Ocean are from the Food and Agriculture Organization (FAO), whose website is http://www.fao.org/fishery/statistics/global-capture-production/query/en, and the time range is 2000–2018. The catch data were downloaded according to the FAOSTAT format, which mainly included aquatic plants, crustaceans, cephalopods, mollusks (except cephalopods), pelagic fishes, demersal fishes, migratory fishes, and other marine fishes. In addition, other marine animals. The data range covers the entire Indian Ocean, including coastal fishing countries and high seas fishing countries.

Let the total catch in the Indian Ocean be the mother sequence X_1, that is, $X_1 = \{x_{1(1)}, x_{0(2)}, \ldots, x_{0(n)}\}$, and let the catches cephalopods, crustaceans, demersal fishes, migratory fishes, and pelagic fishes and the other major species be the subsequences

Table 6.2 The gray correlation coefficients of each category of catch quantum series and the total catch mother series (Lu et al. 2022)

Catch category	Gray relational degree
Demersal marine fish (X_2)	0.91
Crustaceans (X_3)	0.89
Pelagic marine fish (X_4)	0.88
Marine fish NEI (X_5)	0.74
Cephalopods (X_6)	0.70
Mollusks (except cephalopods) mollusks excl. Cephalopods (X_7)	0.56
Freshwater and diadromous fish (X_8)	0.54
Aquatic animals NEI (X_9)	0.52
Aquatic plants (X_{10})	0.49

X_i, i.e., $X_i = \{x_{i\,(1)}, x_{i(2)},\ldots, x_{i(n)}\}$, $i = 1, 2,\ldots, m$. Gray correlation analysis is performed to obtain the main categories that affect the total catch. The averaging method was used for the initial value, and the resolution coefficient was set to 0.5.

The GM (1, *N*) model was used to predict the total catch in the Indian Ocean. The top 5 subsequences with the largest gray correlation degree are selected, and the five GM (1, *N*) prediction models are established according to the degree of gray correlation:

GM (1, 2) model: including the total catch and the maximum correlation value of the catch of the corresponding major species;

GM (1, 3) model: including the total catch and the top 2 correlations. The catch of the corresponding major species;

GM (1, 4) model: including the total catch and the top 3 correlation values. The catch of the corresponding major species;

GM (1, 5) model: including the total catch and the top 4 correlation values of the catch of the corresponding major species;

GM (1, 6) model: including the total catch and the top 5 correlation values of the catch of the corresponding major species;

Five GM (1, N) prediction models were established using the total catches from 2000 to 2016 and the catches of the first five categories, and the average relative error and the gray correlation between the predicted value and the actual value (similarly, averaging was performed, and the resolution coefficient was set to 0.5), and the gray correlation degree with the smallest and largest relative errors was used as the optimal model. The data from 2017 to 2018 were validated. At the same time, the GM (1, 1) model was used to predict the catch of each category in 2019–2025, and then the optimal GM (1, *N*) model was used to predict the total catch in the Indian Ocean from 2019 to 2025.

The gray correlation analysis shows (Table 6.2) that the greatest gray correlation between the total catch (X_1) and the catch of each major species in the Indian Ocean from 2000 to 2016 is the demersal fish (X_2), and its value is the smallest for the aquatic plants (X_{10}), which is 0.52. The top five with the highest degree of gray

Table 6.3 Relevant parameters of the gray GM (1, *N*) model (Lu et al. 2022)

Gray prediction model	Response function	Gray correlation coefficient between the predicted value and the original value
GM (1, 2)	$X_1(t+1) = (9087339.511 - 4.973X_2)\, e^{(-1.604t)} + 4.973X_2$	0.79
GM (1, 3)	$X_1(t+1) = (9087339.511 - 13.874X_2 + 23.564X_3)\, e^{(-0.984t)} + 13.874X_2 - 23.564X_3$	0.70
GM (1, 4)	$X_1(t+1) = (9087339.511 - 15.644X_2 + 17.632X_3 + 2.109X_4)\, e^{(-0.786t)} + 15.644X_2 - 17.632X_3 - 2.109X_4$	0.65
GM (1, 5)	$X_1(t+1) = (9087339.511 - 1.268X_2 - 0.362X_3 - 1.256X_4 - 0.882X_5)\, e^{(-2.061t)} + 1.268X_2 + 0.362X_3 + 1.256X_4 + 0.882X_5$	0.92
GM (1, 6)	$X_1(t+1) = (9087339.511 - 1.057X_2 - 1.345X_3 - 1.083X_4 - 0.915X_5 - 1.071X_6)\, e^{(-1.992t)} + 1.057X_2 + 1.345X_3 + 1.084X_4 + 0.915X_5 + 1.071X_6$	0.92

correlation are demersal fishes (X_2), crustaceans (X_3), pelagic fishes (X_4), other marine fishes (X_5), and cephalopods (X_6).

Statistical analysis showed that the total marine catch in the Indian Ocean from 2000 to 2018 showed a steady growth trend and reached the highest historical production in 2017, which was 12.44 million tons. Among them, pelagic fishes, other marine fishes, demersal fishes, crustaceans, and cephalopods were the main species, and the average catches from 2017 to 2018 were 5,201,000, 2,554,000, 2,541,000, 930,000, and 466,000 tons, respectively. They accounted for 41.98%, 20.61%, 20.51%, 7.51%, and 3.76% of the total catch, respectively.

According to the order of gray correlation value, demersal fish X_2, crustaceans X_3, pelagic fish X_4, other marine fish X_5, and cephalopods X_6 were selected as the factors affecting the total catch X_1. Five GM (1, *N*) models were established in Table 6.3.

The relative errors of the GM (1, 5) and GM (1, 6) models are 1.83% and 1.90%, respectively (Table 6.4). The gray correlation degree between the predicted value and the original value series is 0.92, so the GM (1, 5) and GM (1, 6) models are the optimal prediction models.

According to the optimal models GM (1, 5) and GM (1, 6), the data of 2017 and 2018 were verified, and the average relative errors were 3.78 and 3.43, respectively (Table 6.5), indicating that the accuracy of the model was relatively good.

Using the catch data of demersal fish X_2, crustaceans X_3, pelagic fish X_4, other marine fish X_5, and cephalopods X_6 from 2000 to 2016, the GM (1, 1) model was established (Table 6.6). The 2017–2018 data were used for the test. The catch prediction models for each category basically met the accuracy test requirements. The catches of demersal fishes X_2, crustaceans X_3, pelagic fishes X_4, other marine fishes X_5, and cephalopods X_6 in 2019–2025 are shown in Tables 6.7. According to the GM (1, 5) and GM (1, 6) models, the total catch in the Indian Ocean from 2019 to 2025 can be calculated (Tables 6.8). The total catch will increase between 12.27 and

Table 6.4 Relative error of each GM (1, N) model (Lu et al. 2022)

Year	GM (1, 2)	GM (1, 3)	GM (1, 4)	GM (1, 5)	GM (1, 6)
2001	7.29	14.92	13.67	13.77	13.46
2002	17.44	38.07	41.20	8.49	9.16
2003	2.87	18.36	24.80	2.11	2.79
2004	3.54	5.32	10.99	0.47	0.90
2005	7.17	0.97	0.22	0.00	0.04
2006	7.99	4.95	6.87	0.63	0.80
2007	1.84	0.76	4.28	1.20	1.18
2008	3.03	0.09	5.40	0.08	0.19
2009	2.80	5.36	8.59	0.07	0.05
2010	1.25	17.55	15.76	0.32	0.06
2011	0.80	14.38	9.86	1.05	0.43
2012	7.01	1.04	6.68	0.18	0.44
2013	8.52	8.31	11.80	0.01	0.32
2014	0.75	11.12	21.16	0.36	0.30
2015	3.31	4.90	1.38	0.45	0.16
2016	11.00	28.66	26.33	0.08	0.13
Mean value	5.41	10.92	13.06	1.83	1.90

Table 6.5 Comparison of the predicted and actual values of GM (1, 5) and GM (1, 6) from 2017 to 2018 (Lu et al. 2022)

	2017			2018		
Model	Actual value	Predicted value	Relative error	Actual value	Predicted value	Relative error
GM (1, 5)	12,446,838	11,956,591	3.94	12,332,944	11,901,884	3.50
GM (1, 6)	12,446,838	11,996,265	3.62	12,332,944	11,919,801	3.35

Table 6.6 Prediction models of GM (1, 1) catches (Lu et al. 2022)

Catch category	Response function	Small error probability P	Mean square error ratio C
Demersal fish X_2	$X_2(t+1) = 9246.22\, e^{(0.019t)} - 9046.86$	0.944	0.387
Crustaceans X_3	$X_3(t+1) = 3174.69\, e^{(0.021t)} - 3101.67$	0.944	0.385
Pelagic fish X_4	$X_4(t+1) = 11018.39\, e^{(0.028t)} - 10702.66$	1.000	0.234
Other marine fish X_5	$X_5(t+1) = -18194.06\, e^{(-0.017t)} + 18447.27$	0.650	0.640
Cephalopod X_6	$X_6(t+1) = 268.28\, e^{(0.058t)} - 243.56$	0.944	0.487

Table 6.7 Prediction results of catches of each category from 2019 to 2015 (unit: 10^4 t) (Lu et al. 2022)

Year	Demersal fish	Crustaceans	Pelagic fish	Other marine fish	Cephalopods
2019	257.70	96.88	529.63	226.77	44.83
2020	262.76	98.91	544.89	222.92	47.49
2021	267.92	100.99	560.60	219.14	50.30
2022	273.18	103.11	576.76	215.42	53.28
2023	278.55	105.27	593.38	211.77	56.44
2024	284.02	107.48	610.48	208.18	59.79
2025	289.60	109.74	628.08	204.65	63.33

13.24 million tons in 2021–2025. The main increase in the catch may come from pelagic fish, cephalopods, bottom fish, etc.

6.4.2 Gray Prediction of Shrimp Production in the Bohai Sea

Guo (1992) published "Gray Prediction of Shrimp Yield in the Bohai Sea." In this study, the author used the general multiple linear regression method and the gray system prediction models GM (0, *h*) and GM (1, *h*) to model shrimp production in the Bohai Sea and compared their accuracy.

In this study, X_1 represents the relative yield of *Penaeus chinensis* in the Bohai Sea, and X_2, X_3, and X_4 represent the relative numbers of juvenile shrimp in Bohai Bay, Laizhou Bay, and Liaodong Bay, respectively. The relative yields of prawns and the relative numbers of juveniles are listed in Table 6.9.

We call the data in Tables 6.9 the original sequence, which is denoted as $\left\{X_k^{(0)}(i)\right\}, k=1,\ 2,\ 3,\ 4;\ i=1,\ 2,\ \ldots\ 15$

The general multivariate regression equation established using the data in Table 6.9 is

$$\widehat{X}_1^{(0)}(i)=0.933X_2^{(0)}(i)+0.459X_3^{(0)}(i)+0.366X_4^{(0)}(i)-1.694$$

Second, the gray static multivariate model GM (0, 4) is established:

$$\widehat{X}_1^{(1)}(i)=0.987X_2^{(1)}(i)+0.493X_3^{(1)}(i)+0.297X_4^{(1)}(i)-32.509$$

The gray dynamic multivariate model GM (1, 4) is

$$\frac{dX_1^{(1)}}{dt}+1.731X_1^{(1)}=1.778X_2^{(1)}+0.763X_3^{(1)}+0.568X_4^{(1)}$$

The corresponding time-corresponding equation is

Table 6.8 Forecast results of the total catch in the Indian Ocean from 2019 to 2025 Unit: t (Lu et al. 2022)

Year	2019	2020	2021	2022	2023	2024	2025
GM (1, 5)	11,865,349	12,065,390	12,272,166	12,485,569	12,705,897	12,933,212	13,167,845
GM (1, 6)	11,882,322	12,089,118	12,303,560	12,525,639	12,755,711	12,994,002	13,240,790

Table 6.9 Relevant data of *Penaeus chinensis* in the Bohai Sea (Guo 1992)

Year	Relative number of juvenile shrimp			Relative yield of prawns	Forecast value		
	Bohai Bay	Laizhou Bay	Liaodong Bay		Regression	GM (0, *h*)	GM (1, *h*)
1969	119	48	1	100.0	133.5	108.9	100.0
1972	20	157	16	100.5	100.6	101.9	114
1973	66	243	100	236.8	216.9	214.6	221.4
1974	13.9	165	251	313.4	301.7	293.0	305
1975	100	314	114	254.0	288.9	287.3	281.1
1976	64	37	39	87.4	90.6	93.0	95.6
1977	158	123	44	212.7	222.3	229.6	231.1
1978	163	223	42	320.0	276.3	283.3	279.5
1979	305	191	133	404.9	426.3	434.6	441.1
1980	176	276	2	313.2	300.0	310.3	303.1
1981	119	117	46	205.6	184.2	188.8	188.9
1982	20	61	23	58.2	55.6	55.6	55.0
1983	91	72	25	147.1	128.1	132.7	133.4
1984	48	21	27	53.5	63.4	65.7	67.4
1985	24	323	33	174.3	192.8	192.7	177.8

$$\widehat{X}_1^{(1)}(i) = \left(100 - 1.027X_2^{()}(i) - 0.441X_3^{(1)}(i) - 0.325X_4^{(1)}(i)\right)e^{-1.72t(i-1)}$$
$$+1.027X_2^{(1)}(i) + 0.441X_3^{(1)}(i) + 0.328X_4^{(1)}(i)$$

The above three models were used to predict the relative yields of prawns in Table 6.8, and the results are listed in Table 6.8. The absolute and relative deviations of the predicted values of the three models relative to the original series values are listed in Tables 6.10.

The analysis shows that, under the condition that the relative number of juvenile shrimp in each bay is known, the multivariate gray model established by the gray system method has higher prediction accuracy and higher reliability than the traditional regression model. It can also be seen from the prediction results that GM (0, *h*) and GM (1, *h*) have similar prediction accuracies and can be used to predict shrimp in each bay.

6.4.3 Gray Prediction of Fishery Human Resources

Chen and Zhou (2001) published "Analysis and Prediction of the Human Resources Structure of China's Marine Fisheries." Based on the statistical data of China's marine fishery human resources between 1990 and 1998, the study analyzed China's 1990–1998 period using gray correlation and gray prediction methods. The

Table 6.10 Error analysis of the three models (Guo 1992)

Year	Regression model		GM (0, *h*)		GM (1, *h*)	
	Absolute error	Relative error	Absolute error	Relative error	Absolute error	Relative error
1969	33.5	33.5	8.9	8.9	0	0
1972	0.1	0.1	1.4	1.4	13.6	13.6
1973	19.9	8.4	22.2	9.4	13.4	6.5
1974	11.7	3.7	20.4	6.5	8.3	2.7
1975	34.9	13.7	33.3	13.1	27.1	10.7
1976	3.2	3.7	5.6	6.4	8.2	9.3
1977	10.1	4.7	16.9	8.0	18.4	8.6
1978	43.7	13.7	36.7	11.5	40.5	12.7
1979	21.4	5.3	29.7	7.3	36.2	8.9
1980	13.2	4.2	2.9	0.9	10.1	3.2
1981	21.4	10.4	16.8	8.2	16.7	8.1
1982	2.6	4.5	1.6	2.7	3.2	5.6
1983	19.0	12.9	14.4	9.8	13.7	9.3
1984	9.9	18.5	12.2	22.9	13.9	26.0
1985	18.5	10.6	18.3	10.5	3.4	2.0

Table 6.11 Statistics of China's marine fishery labor from 1990 to 1998. Unit: person (Chen and Zhou 2001)

Year	Fishery labor X_0	Fishing labor X_1	Farming labor X_2	Service labor X_3	Part-time labor X_4
1990	2,080,537	960,800	257,119	173,758	688,860
1991	2,167,621	971,668	276,880	216,354	702,719
1992	2,240,263	1,023,730	296,592	204,911	715,030
1993	2,329,479	1,046,095	355,943	211,100	716,341
1994	2,386,469	1,052,384	365,933	239,101	729,051
1995	2,514,682	1,099,454	398,715	244,888	771,625
1996	2,526,353	1,167,362	319,486	232,816	806,689
1997	2,681,563	1,193,838	459,177	269,779	758,769
1998	2,711,360	1,185,079	488,706	268,956	768,619

composition of the marine fishery labor and its changes and the gray forecast of the development trend of China's marine fishery labor from 2000 to 2005. The original data are shown in Table 6.11.

The raw data in Tables 6.10 were subjected to initial transformation to calculate the relative gray correlation between fishery labor and fishing labor, farming labor, service labor, and part-time labor, and the values are $\gamma_{01} = 0.9029$, $\gamma_{02} = 0.6477$, $\gamma_{03} = 0.6719$, and $\gamma_{04} = 0.8057$, respectively.

At the same time, each of the original sequences in Tables 6.10 was treated with the zero value of the starting point, and the absolute gray correlations ε between the fishery labor and fishing labor, farming labor, service labor, and part-time labor were calculated as follows: $\varepsilon_{01} = 0.68110$, $\varepsilon_{02} = 0.65178$, $\varepsilon_{03} = 0.58665$, $\varepsilon_{04} = 0.58046$.

If $\theta = 0.5$, then the gray comprehensive degree is $R = 0.5\gamma + 0.5\varepsilon$; then, we can obtain:

$$R_{01} = 0.7920, R_{02} = 0.6497, R_{03} = 0.6293, R_{04} = 0.6931$$

6.4.3.1 Gray Prediction of Fishery Labor

The accumulative number is generated once for the original data sequence X_0, and the accumulative sequence is obtained as:

$$X_0{}^{(1)} = (2080537, 4248158, 6488421, 8817900, 11204369, 13719051, 16245404, 18926967, 21638327).$$

According to the cumulative sequence $X_0{}^{(1)}$, the adjacent mean sequence is generated to obtain the adjacent mean sequence.

$$Z_0{}^{(1)} = (3164347.5, 5368290, 7653160.5, 10011135, 12461710, 14982228, 17586186, 20282647)$$

Construct matrices B and Y, respectively, and obtain the parameter sequence

$$\widehat{a} = (a, b)^T = \left(B^T B^{-1} B^T Y\right) = \begin{vmatrix} -0.03274 \\ 2070175 \end{vmatrix}$$

Then, the GM (1, 1) model of marine fishery labor is

$$\frac{dx^{(1)}}{dt} - 0.03274x^{(1)} = 2070175$$

The corresponding time response is

$$\widehat{x}^{(1)}(k+1) = \left(x^{(0)}(1) - \frac{b}{a}\right)e^{-ak} + \frac{b}{a} = 65303597\,e^{-0.03274k} - 63223060$$

The simulation value was calculated using the time response equation

Table 6.12 The error test table (Chen and Zhou 2001)

Serial number	Actual data	Simulation data	Residual	Relative error %
2	2,167,621	2,173,693	−6072	0.28
3	2,240,263	2,246,047	−5784	0.26
4	2,329,479	2,320,809	8670	0.37
5	2,386,469	2,398,059	−11,590	0.49
6	2,514,682	2,477,881	36,801	1.46
7	2,526,353	2,560,359	−34,006	1.35
8	2,681,563	2,645,583	35,980	1.34
9	2,711,360	2,733,644	−22,284	0.82

$$\widehat{x}^{(1)} = (2080537, 4254230, 6500277, 8821086, 11219145, 13697025, 16257385, 18902968, 21636612)$$

Solving the simulation value of $X^{(0)}\,\widehat{x}^{(0)}$, we have

$$\widehat{x}^{(0)} = (2080537, 2173693, 2246047, 2320809, 2398059, 2477881, 2560359, 2645583, 2733644)$$

The original actual data and the simulated data were compared, and the residuals and relative errors were obtained. The results are shown in Table 6.12.

Average relative error $\Delta = \frac{1}{8}\sum_{k=2}^{9}\Delta_k = 0.80\% < 0.01$.

The analysis of the average relative error indicates that the accuracy of the model reaches the first level, which meets the requirements of prediction.

The development trend of marine fishery labor from 2000 to 2005 was predicted, and the simulation value $\widehat{x}^{(1)}$ was obtained:

$$\begin{aligned}\widehat{x}^{(1)} &= \left(\widehat{x}^{(1)}\,(2000), \widehat{x}^{(1)}\,(2001), \ldots, \widehat{x}^{(1)}\,(2004), \widehat{x}^{(1)}\,(2005)\right)\\ &= \left(\widehat{x}^{(1)}\,(11), \widehat{x}^{(1)}\,(12), \ldots, \widehat{x}^{(1)}\,(15), \widehat{x}^{(1)}\,(16)\right)\\ &= (27379905, 30395712, \ldots, 40058915, 43496755)\end{aligned}$$

The predicted value $\widehat{x}^{(0)}$ of $X^{(0)}$ from 2000 to 2005 is obtained by reduction.

$$\begin{aligned}\widehat{x}^{(0)} &= \left(\widehat{x}^{(0)}\,(2000), \widehat{x}^{(0)}\,(2001), \ldots, \widehat{x}^{(0)}\,(2005)\right)\\ &= (2918657, 3015807, 3116191, 3219917, 3327095, 3437840)\end{aligned}$$

6.4.3.2 Gray Prediction of Fishing Labor

Using the sequence of marine fishing labor force X_1, the GM (1, 1) model is obtained through the above calculation:

$$\frac{dx^{(1)}}{dt} - 0.02994x^{(1)} = 938014$$

The corresponding time response is

$$\hat{x}^{(1)}(k+1) = \left(x^{(0)}(1) - \frac{b}{a}\right)e^{-ak} + \frac{b}{a} = 32291627e^{-0.02994k} - 31330827$$

The simulation data were obtained, and the error test was performed. The test results are shown in Table 6.13.

Then, the average relative error is $\Delta = \frac{1}{8}\sum_{k=2}^{9} \Delta_k = 1.31\% < 0.05$, and the error accuracy reaches the second level, which meets the requirements of prediction.

The development trend of marine fishing labor from 2000 to 2005 was predicted, and the predicted value $\hat{x}^{(0)}$ of $X^{(0)}$ from 2000 to 2005 was recovered:

$$\hat{x}^{(0)} = \left(\hat{x}^{(0)}\ (2000),\ \hat{x}^{(0)}\ (2001), \ldots,\ \hat{x}^{(0)}\ (2004),\ \hat{x}^{(0)}\ (2005)\right)$$
$$= (1284890, 1323940, 1364176, 1405636, 1448356, 1492374)$$

6.4.3.3 Gray Prediction of the Farming Labor

Using the sequence of the farming labor X_2, the GM (1, 1) model is obtained through the above calculation:

Table 6.13 The error test table (Chen and Zhou 2001)

Serial number	Actual data	Simulation data	Residual	Relative error%
2	971,668	981,397	−9729	0.10
3	1,023,730	1,011,223	12,507	1.22
4	1,046,095	1,041,956	4139	0.40
5	1,052,384	1,073,623	−21,239	2.02
6	1,099,454	1,106,252	−6798	0.62
7	1,167,362	1,139,873	27,489	2.35
8	1,193,838	1,174,516	19,322	1.62
9	1,185,079	1,210,211	−25,132	2.12

Table 6.14 The error test table (Chen and Zhou 2001)

Serial number	Actual data	Simulation data	Residual	Relative error%
2	276,880	283,232	−6352	2.29
3	296,592	304,477	−7885	2.66
4	355,943	327,315	28,628	8.04
5	365,933	351,867	14,066	3.84
6	398,715	378,260	20,455	5.13
7	319,486	406,632	−87,146	27.28
8	459,177	437,133	22,044	4.80
9	488,706	469,921	18,785	3.84

$$\frac{dx^{(1)}}{dt} - 0.07233x^{(1)} = 254516$$

The corresponding time response is

$$\widehat{x}^{(1)}(k+1) = \left(x^{(0)}(1) - \frac{b}{a}\right)e^{-ak} + \frac{b}{a} = 3776012e^{-0.07233k} - 3518893$$

The simulated data were obtained and tested for errors. The test results are shown in Table 6.14.

Then, the average relative error $\Delta = \frac{1}{8}\sum_{k=2}^{9}\Delta_k = 7.24\% < 0.10$, and the error accuracy reaches the third level, which can basically meet the requirements of prediction.

The development trend of farming labor from 2000 to 2005 was predicted, and the predicted value $\widehat{x}^{(0)}$ of $X^{(0)}$ from 2000 to 2005 was restored:

$$\widehat{x}^{(0)} = \left(\widehat{x}^{(0)}\ (2000),\ \widehat{x}^{(0)}\ (2001), \ldots,\ \widehat{x}^{(0)}\ (2004),\ \widehat{x}^{(0)}\ (2005)\right)$$
$$= (543061, 583795, 627585, 674659, 725264, 779665)$$

6.4.3.4 Gray Prediction of Service Labor

Using the sequence of service labor X_3, the GM (1, 1) model is obtained through the above calculation:

$$\frac{dx^{(1)}}{dt} - 0.03892x^{(1)} = 194347$$

The corresponding time response is

Table 6.15 The error test table (Chen and Zhou 2001)

Serial number	Actual data	Simulation data	Residual	Relative error%
2	216,354	205,074	11,280	5.21
3	204,911	213,212	−8301	4.05
4	211,100	221,673	−10,573	5.01
5	239,101	230,470	8631	3.61
6	244,888	239,616	5272	2.15
7	232,816	249,125	−16,309	7.01
8	269,779	259,011	10,768	3.99
9	268,956	269,290	−334	0.12

$$\widehat{x}^{(1)}(k+1)=\left(x^{(0)}(1)-\frac{b}{a}\right)e^{-ak}+\frac{b}{a}=5167668e^{-0.03892k}-4993910$$

The simulated data were obtained and tested for errors. The test results are shown in Table 6.15.

Then, the average relative error $\Delta=\frac{1}{8}\sum_{k=2}^{9}\Delta_k=3.89\%<0.05$, and the error accuracy reaches the second level, which meets the requirements of prediction.

The development trend of service labor from 2000 to 2005 was predicted, and the predicted value $\widehat{x}^{(0)}$ of $X^{(0)}$ from 2000 to 2005 was restored.

$$\begin{aligned}\widehat{x}^{(0)}&=\left(\widehat{x}^{(0)}(2000),\ \widehat{x}^{(0)}(2001),\ \ldots,\ \widehat{x}^{(0)}(2004),\ \widehat{x}^{(0)}(2005)\right)\\&=(291087,302639,314649,327135,340117,353614)\end{aligned}$$

6.4.3.5 Gray Prediction of Part-Time Labor

The GM (1, 1) model is obtained through the above calculation using the sequence of the part-time marine labor force X_4:

$$\frac{dx^{(1)}}{dt}-0.01570x^{(1)}=689422$$

The corresponding time response is

$$\widehat{x}^{(1)}(k+1)=\left(x^{(0)}(1)-\frac{b}{a}\right)e^{-ak}+\frac{b}{a}=44610062e^{-0.01570k}-43921202$$

The model was used to obtain the simulated data, and the error test was performed. The test results are shown in Table 6.16.

Table 6.16 The error test table (Chen and Zhou 2001)

Serial number	Actual data	Simulation data	Residual	Relative error%
2	702,719	705,759	−3040	0.43
3	715,030	716,924	−1894	0.26
4	716,341	728,267	−11,926	1.66
5	729,051	739,788	−10,737	1.47
6	771,625	751,492	20,133	2.61
7	806,689	763,381	43,308	5.37
8	758,769	775,458	−16,689	2.20
9	768,619	787,727	−19,108	2.49

Then, the average relative error $\Delta = \frac{1}{8}\sum_{k=2}^{9}\Delta_k = 2.06\% < 0.05$, and the error accuracy reaches the second level, which meets the prediction requirements.

The development trend of part-time labor from 2000 to 2005 was predicted, and the predicted value $\widehat{x}^{(0)}$ of $X^{(0)}$ from 2000 to 2005 was restored:

$$\widehat{x}^{(0)} = \left(\widehat{x}^{(0)}(2000),\ \widehat{x}^{(0)}(2001),\ \ldots, \widehat{x}^{(0)}(2004),\ \widehat{x}^{(0)}(2005)\right)$$
$$= (812848, 825708, 838771, 852041, 865521, 879214)$$

The studies have shown that the order of gray comprehensive correlation from high to low is $R_{01} > R_{04} > R_{02} > R_{03}$. The factors that have the greatest impact on marine fishery labor are fishing labor, followed by part-time labor, farming labor, and farming labor. The study suggests that labor marine fishing is the largest factor affecting labor in marine fisheries. This indicates that in the current human resource structure of marine fisheries in China, the number of people engaged in fishing is large, while the number of people engaged in farming is small. The overexploitation of offshore fishery resources will result in excessive fishing capacity. This structure of human resources is extremely detrimental to the sustainable development of China's offshore fishery resources.

Through the establishment of the gray model GM (1, 1), the development trends of marine fisheries, marine fishing, marine farming, service labor, and part-time labor in 2000–2005 were predicted. By 2000–2005, the number of laborers engaged in marine fisheries in China will reach 2.919–3.438 million, the number of laborers in marine fishing will be 1.285–1.492 million, and the number of laborers in marine farming will be 543,000–779,000. The number of people working part-time in the marine industry is between 813,000 and 879,000.

6.4.4 Application of the Gray System in Fishery Forecasting

Fishery forecasting is the main content of fishery oceanography. Fishery forecasting refers to the forecast of various factors of aquatic living resource conditions in a certain period of time and within a certain range of marine waters in the future, such as fishing season, fishing grounds, quantity and quality of fish stocks, and possible catches. The basis of fishery forecasting is the relationship between fish behavior and biological conditions and environmental conditions and their patterns, as well as various fishing and sea condition data obtained from various real-time fisheries surveys, such as catches, resource conditions, and marine environments. The main task of fishery forecasting is to predict the fishing grounds, fishing season, and possible catch. In recent years, the gray system has been used for fishery forecasting and has achieved good results.

6.4.4.1 Gray Prediction of the Peak Fishing Season of *Ommastrephes bartramii*

Understanding the fishing season characteristics of *Ommastrephes bartramii* in the northern Pacific Ocean and predicting the peak fishing season is one of the important parts of fishery forecasting, which is conducive to the scientific production of fishery enterprises, the rationalization of the operation time, the saving of costs, and the improvement of fishing efficiency. Gray waveform prediction is a prediction method for periodic fluctuation series in the gray system. Based on the GM (1, 1) model group, it can perform medium- and long-term predictions for time series with large fluctuation amplitudes and small amounts of data. The squid, *Ommastrephes bartramii*, in the northern Pacific Ocean is an annual species, its fishing season is the annual cycle, and the annual variation is drastic. It is difficult to use the traditional statistical analysis method to carry out better prediction of the changes in its fishing season. To this end, Xie and Chen (2021a) used fishery production statistical data from 2013 to 2017 to establish a series of fishing season dates and used the gray waveform prediction model to predict the peak fishing season of *Ommastrephes bartramii*.

6.4.4.1.1 Data and Study Methods

The statistical data of fishery production are from China squid jigging vessels. The time period is from 2013 to 2017, and the spatial range is 35°–45°N and 140°–179° E. The fishing data include date, longitude, latitude, and daily yield. The spatial resolution is 1° × 1°. The catch per unit fishing effort (CPUE) was used to characterize the abundance index of *Ommastrephes bartramii*.

Because the annual yield is affected by the marine climate environment and fluctuates greatly, the quartile of the $CPUE_{day}$ sequence ($Q_1 - Q_3$) is calculated based on the $CPUE_{day}$ sequence of the current year in each year. The third quantile Q_3 of the sequence ($CPUE_{day}$ value greater than 75%) was defined as the high-yield $CPUE_{day}$ value. If there is a high-yield $CPUE_{day}$ value for more than 3 consecutive days, the first day of these 3 days is called the start of the peak fishing season. If there is no high-yield $CPUE_{day}$ value for more than 3 consecutive days, the first day of these 3 days is called the end of the peak fishing season.

The gray waveform prediction model was used to predict the fishing season of *Ommastrephes bartramii* in the North Pacific Ocean. The study method is as follows:

1. Data selection: According to the division results of the peak fishing season, the sequence number of the beginning of the peak fishing season is found from the fishing season time to form the peak fishing season date series. Suppose the date sequence of the fishing season $X = (x\ (1), x\ (2)..., x\ (n))$, then the n-segment polyline graph of the sequence X is called $x_n = x\ (n) + (t - n)\ [x\ (n + 1) - x\ (n)]$, i.e., the sequence $X = \{x_n = x\ (n) + (t - n)\ [x\ (n + 1) - x\ (n)] \mid n = 1, 2..., m - 1\}$.
2. Selection of contour lines. Suppose $\sigma_{\max} = \max\limits_{1 \le n \le m} \{x(n)\}$ and $\sigma_{\min} = \min\limits_{1 \le n \le m} \{x(n)\}$, we select $s + 1$ thresholds $\xi_0, \xi_1, \ldots, \xi_s$ between $\sigma_{\max}$ and $\sigma_{\min}$, so that it satisfies $\forall_{\xi} \in [\sigma_{\min}, \sigma_{\max}]$, and

$$\xi_0 = \sigma_{\min}, \quad \ldots, \quad \xi_i = \frac{i}{s}(\sigma_{\max} - \sigma_{\min}) + \sigma_{\min}, \quad \ldots, \quad \xi_s = \sigma_{\max};$$

   Then, $X = \xi_i\ (i = 1, 2 ..., s)$ is the $s + 1$ contour line with equal intervals of the line graph X.
3. Selection of contour time sequence. The contour line $X = \xi_i$ intersects with the line graph of the peak fishing season's prosperity. If there is a contour on the K broken line of X, the coordinates are $\left(k + \frac{\xi - x(k)}{x(k+1) - x(k)}, \xi\right)$. Let $q(k) = k + \frac{\xi - x(k)}{x(k+1) - x(k)}$ $(k = 1, 2 ..., m)$; then, $Q_i = (q\ (1), q\ (2) ..., q\ (m))$ $(k = 0, 1 ..., s)$ is the time sequence of the contour.
4. Establish the GM (1, 1) model for the time sequence of the contour. For each contour line, the contour time sequence $Q_i = (q\ (1), q\ (2) ..., q\ (m))$ $(k = 0, 1..., s)$ is used to establish a GM (1, 1) model to obtain the predicted values. See Chap. 5 for the calculation and testing of the GM (1, 1) model.

6.4.4.1.2 Research Results

Analysis of Fishing Season Characteristics

In 2013–2017, the earliest fishing season of *Ommastrephes bartramii* in the northern Pacific Ocean was May 12 (2017), the latest was December 31 (2014–2016), and the peak fishing season was June–November (Fig. 6.1). The range of total fishing operation days was 194–224 days, with an average of 213 days; the daily unit

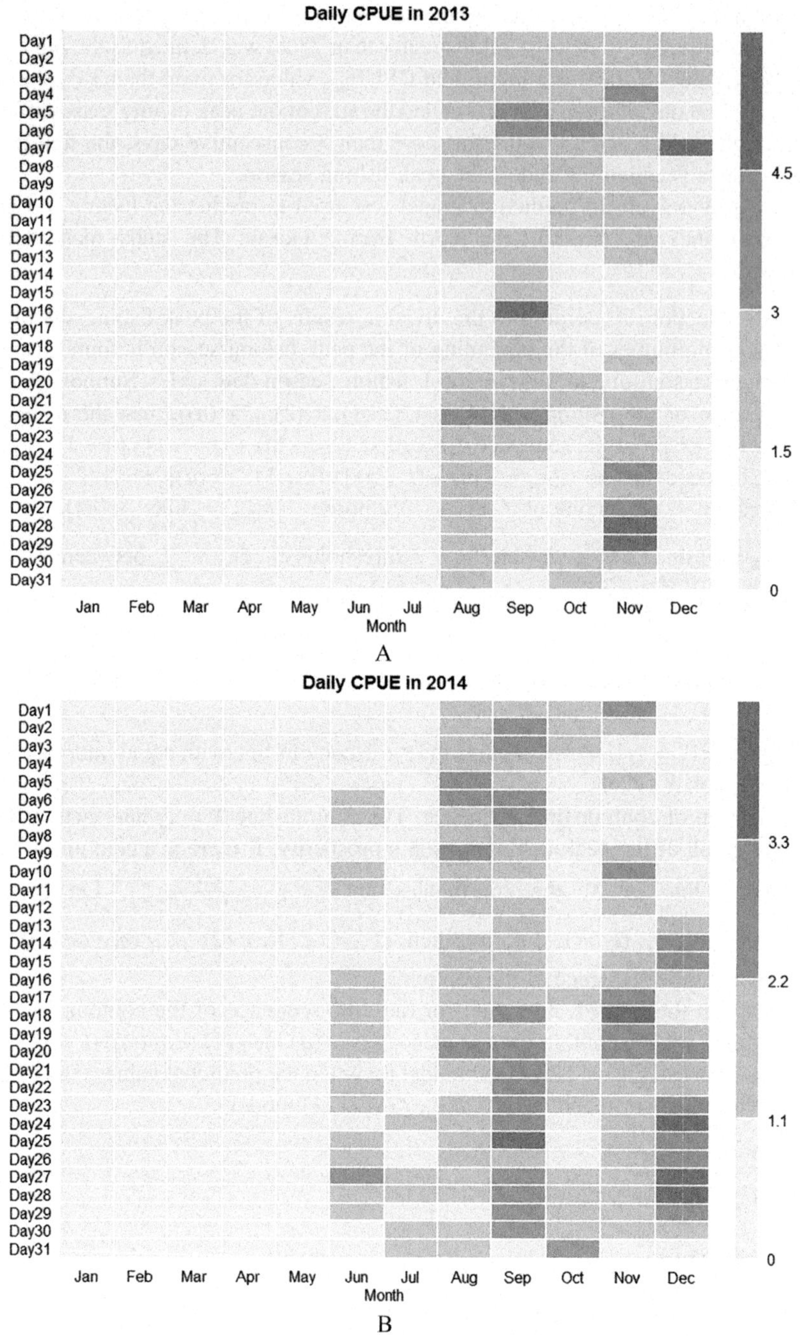

Fig. 6.1 Distribution of daily catch per unit of fishing effort (A–E) for *Ommastrephes bartramii* in the northern Pacific Ocean during 2013–2017 (Xie and Chen 2021a)

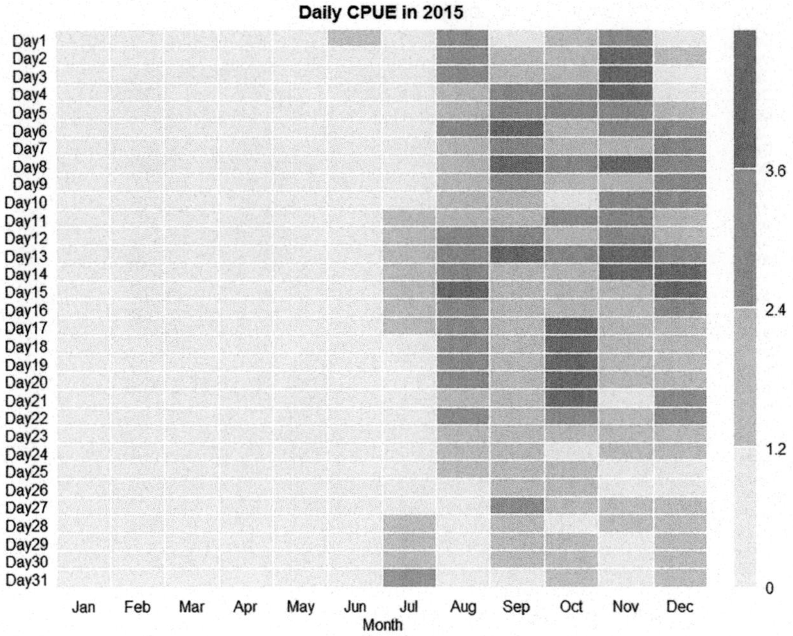

C

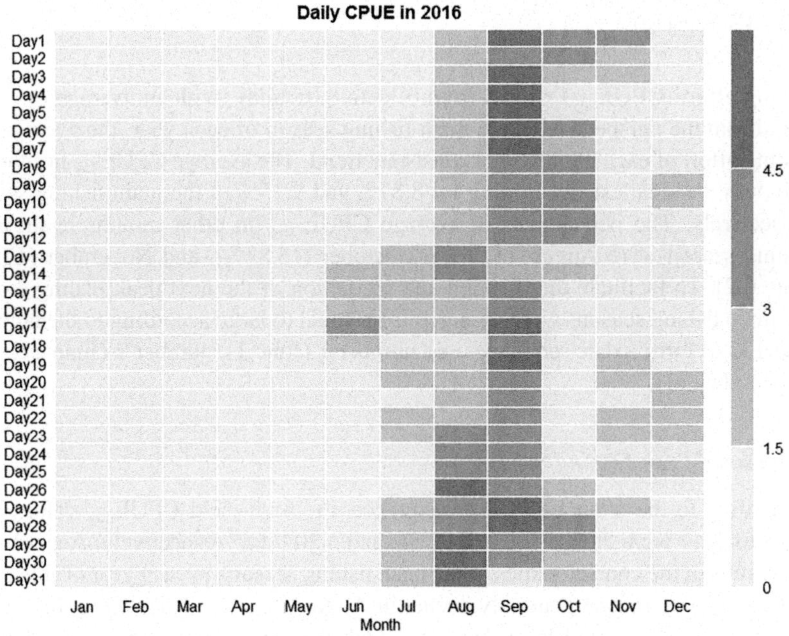

D

Fig. 6.1 (continued)

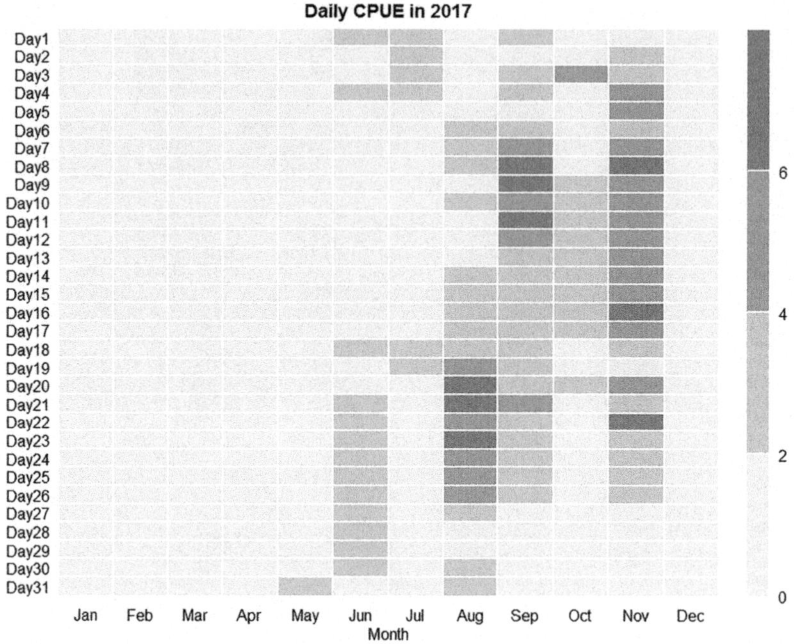

E

Fig. 6.1 (continued)

fishing effort ($CPUE_{day}$) ranged from 0.13 to 8.05 t/day, with an average of 1.82 t/day. From the perspective of the main fishing season of each year, the average unit fishing effort of each month has a consistent trend. The average $CPUE_{day}$ in June and July was the smallest, accounting for 8.87% and 9.15% of the main fishing season, respectively. The percentages of average $CPUE_{day}$ in other months in the main fishing season were August (17.81%), October (13.82%), and November (17.4%) (Fig. 6.2). To facilitate the division and prediction of the next peak fishing season, the main fishing seasons were arranged from small to large according to the numbers 1 to 183 to form a date sequence, i.e., number 1 (June 1), number 2 (June 2),...,183 (November 31).

Analysis of Peak Fishing Season

According to the annual $CPUE_{day}$ sequence, the high-yield $CPUE_{day}$ values from 2013 to 2017 were 2.07, 1.91, 2.51, 2.86, and 3.20 t/day, respectively. According to the results of the characteristics of the peak fishing seasons in each year (Table 6.17), the first fishing season generally occurs in August, and it is basically in the second 10 days of August, and the subsequent peak fishing seasons occur in September, October, and November. Except for only one peak fishing season in 2016, there were at least four peak fishing seasons in all other years, and the highest peak fishing

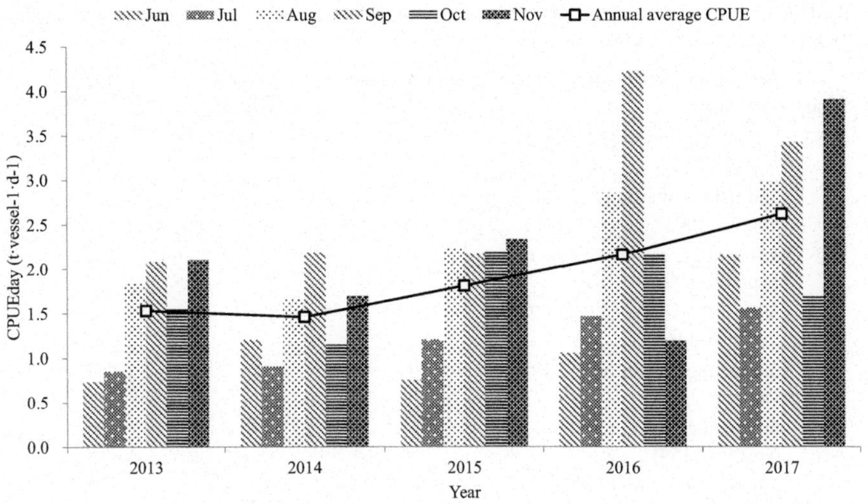

Fig. 6.2 Daily catch per unit of fishing effort (A–E) for *Ommastrephes bartramii* in the northern Pacific Ocean from June to November 2013–2017 (Xie and Chen 2021a)

season occurred in 2013, with seven peak fishing seasons. The number of days in each peak fishing season is different. The minimum number of days is 3 (the first fishing season in 2013 and 2015), the maximum is 51 days (the first fishing season in 2016), and the average number of days is 10 days. The average $CPUE_{day}$ in each fishing season was above 1.99 t/day, and the highest $CPUE_{day}$ was 8.05 t/day (the fourth fishing season in 2017). In general (except for 2017), the average $CPUE_{day}$ tends to increase with increasing time in each peak season of each year.

According to the date series of the peak fishing season from 2013 to 2015, 10 contour lines X ($X_0 - X_9$) are divided. Since at least four data sets are required for GM (1, 1) modeling, GM (1, 1) modeling is performed on only 7 sets of contour sequences ($X_2 - X_8$). From the fitting results of the model (Table 6.18), the average relative error is within 11.58%, and the fitting effect of the GM (1, 1) model with contour line X_2 is the best, which is 1.74%. From the perspective of the relevant parameters of the model, the probability of small error P is 1.00 (>0.95); the variance ratio of the contour line X_7 and the contour line X_8 model is $C < 0.50$, and the variance ratios of other models are all consistent with $C < 0.35$. The accuracy of the model is grade I and grade II. From the perspective of the development coefficient a of the model, all models can be used for medium- and long-term prediction ($-a \leq 0.3$).

According to the results of the GM (1, 1) prediction model in the peak fishing season (Table 6.19), the fitting effect was good in the peak fishing season, except for the relatively large error in the fourth fishing season in 2014 (relative error of 49.01%). The relative error of the fitting during the peak fishing season was within

Table 6.17 Characteristics of peak fishing season of *Ommastrephes bartramii* in the northern Pacific Ocean (Xie and Chen 2021a)

Year	Peak fishing season (month-day/date sequence)	Numbers of days	Average $CPUE_{day}$	Highest $CPUE_{day}$
2013	1st fishing season (22 Aug-24 Aug/83–85)	3	2.57	3.04
	2nd fishing season (3 Sep-8 Sep/95–100)	6	2.58	3.13
	3rd fishing season (12 Sep-17 Sep/104–109)	6	2.76	4.61
	4th fishing season (21 Sep-24 Sep/113–116)	4	2.69	3.57
	5th fishing season (4 Oct-12 Oct/126–134)	9	2.6	3.41
	6th fishing season (3 Nov-6 Nov/156–159)	4	2.61	3.21
	7th fishing season (25 Nov-30 Nov/178–183)	6	3.65	5.07
2014	1st fishing season (4 Aug-10 Aug/65–71)	7	1.99	2.31
	2nd fishing season (17 Aug-20 Aug/78–81)	4	2.14	2.63
	3rd fishing season (1 Sep-8 Sep/93–100)	8	2.23	2.49
	4th fishing season (16 Sep-2 Oct/108–124)	17	2.43	3.54
	5th fishing season (17 Nov-21 Nov/170–174)	5	2.78	3.75
2015	1st fishing season (1 Aug-3 Aug/62–64)	3	2.89	3.46
	2nd fishing season (12 Aug-20 Aug/73–81)	9	3.11	3.71
	3rd fishing season (4 Sep-13 Sep/96–105)	10	3.11	4.29
	4th fishing season (17 Nov-22 Nov/139–145)	6	3.59	3.92
	5th fishing season (1 Nov-14 Nov/154–168)	14	3.21	4.06
2016	1st fishing season (23 Aug-12 Oct/84–134)	51	4.01	5.84
2017	1st fishing season (19 Aug-26Aug/80–87)	8	5.23	6.65
	2nd fishing season (6 Sep-15 Sep/98–107)	10	5.17	6.34
	3rd fishing season (19 Sep-22 Sep/111–114)	4	3.87	4.66
	4th fishing season (2 Nov-25 Nov/155–178)	24	4.56	8.05

Table 6.18 Fitting results and relevant parameters of GM (1, 1) models with different contours (Xie and Chen 2021a)

Contours X	The fitting contour sequence results by GM (1, 1) model					Average relative error	Variance ratio C	Development coefficient a	Small error probability P
$X_2 = 87.78$	Contour time sequence	1.40	7.80	9.65	12.76	1.74	0.04	−0.25	1.00
	GM (1, 1) model	1.40	7.62	9.78	12.57				
$X_3 = 100.67$	Contour time sequence	2.63	7.68	10.51	12.64	2.72	0.08	−0.24	1.00
	GM (1, 1) model	2.63	7.91	10.03	12.72				
$X_4 = 113.56$	Contour time sequence	4.04	7.57	11.09	12.52	5.77	0.19	−0.23	1.00
	GM (1, 1) model	4.04	8.09	10.17	12.80				
$X_5 = 126.44$	Contour time sequence	5.01	7.46	11.30	12.40	7.21	0.26	−0.23	1.00
	GM (1, 1) model	5.01	8.11	10.18	12.77				
$X_6 = 139.33$	Contour time sequence	5.44	7.34	11.51	12.28	8.66	0.33	−0.22	1.00
	GM (1, 1) model	5.44	8.13	10.18	12.74				
$X_7 = 152.22$	Contour time sequence	5.87	7.23	11.71	12.16	10.12	0.39	−0.22	1.00
	GM (1, 1) model	5.87	8.15	10.18	12.71				
$X_8 = 165.11$	Contour time sequence	6.41	7.11	11.92	12.05	11.58	0.47	−0.22	1.00
	GM (1, 1) model	6.41	8.17	10.18	12.68				

Table 6.19 Relative errors of the GM (1, 1) prediction model during the peak fishing season (Xie and Chen 2021a)

Peak fishing season	Actual value	Predicted value	Relative error
1st fishing season in 2014	65.00	71.07	9.34
2nd fishing season in 2014	78.00	77.90	0.12
3rd fishing season in 2014	93.00	97.54	4.88
4th fishing season in 2014	108.00	160.93	49.01
5th fishing season in 2014	170.00	170.51	0.30
2nd fishing season in 2015	93.00	98.75	6.18
3rd fishing season in 2015	108.00	135.50	25.46

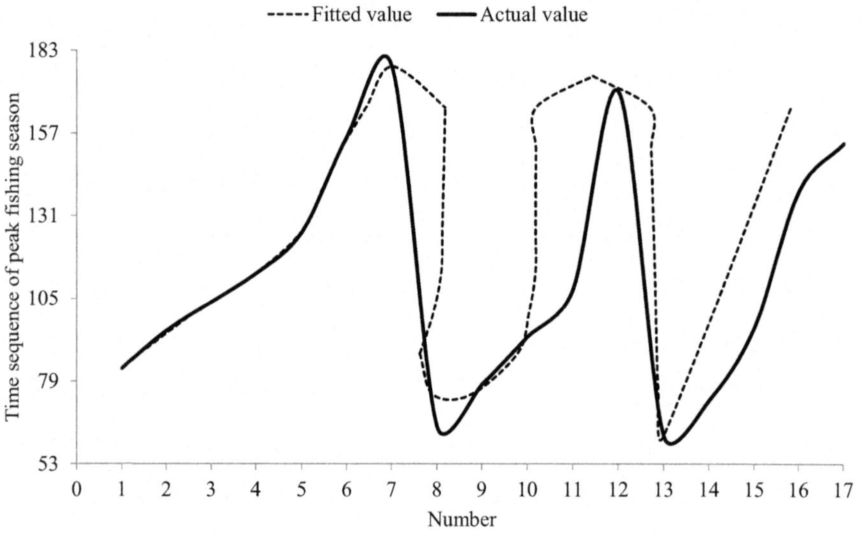

Fig. 6.3 Comparison of the predicted values and the actual values of *Ommastrephes bartramii* based on the main fishing season forecasting model (Xie and Chen 2021a)

9.34%, and the average relative error was 12.73%. The average relative error of the verification during the peak fishing season in 2015 was 15.82%.

In this study, we used the gray waveform prediction method and established the GM (1, 1) model to predict the peak fishing season of Ommatrephes bartramii. In terms of the relationship between the predicted values and the actual values (Fig. 6.3), the variation trend of the CPUE is basically the same. From the perspective of the parameters of the prediction model, the model has good accuracy (Table 6.18) and can be used for medium- and long-term prediction. However, this study only considered the changes in fishery production data and did not include the climatic and marine environmental factors that affect the changes in the abundance of Ommatrephes bartramii.

6.4.4.2 Establishment of a Gray Prediction Model for the Abundance Index of *Ommastrephes bartramii*

The prediction of the abundance index is also an important part of fishery forecasting. Constructing a reasonable gray forecasting model is the basis of scientific forecasting. To this end, Xie and Chen (2021b) first established a GM (1, 1) model group for the abundance index sequences of *Ommastrephes bartramii* with different time-series lengths and selected the CPUE sequence with the smallest relative error and variance as the mother sequence. Second, the gray correlations between the mother sequence and Pacific interdecadal oscillation index (PDO), mean sea surface temperature (SGSST) of the spawning field, mean sea surface temperature of the fattening field (FGSST), mean chlorophyll concentration of the spawning field (SGC), and mean chlorophyll concentration of the fattening field (FGC) are used to evaluate the effect of environmental factors on the abundance index of *Ommastrephes bartramii*. Based on the evaluation results, six gray prediction models with different orders, including GM (0, *N*) and GM (1, *N*), were established. The model with the smallest error was selected as the best model for predicting the abundance index of *Ommastrephes bartramii*, which can provide a basis for the scientific production of squid fishing vessels in the North Pacific Ocean.

6.4.4.2.1 Research Data and Methods

The statistical data of fishery production are from the Chinese squid jigging vessels. The time period is from 1998 to 2016, and the spatial range is 35°–45°N and 140°–179°E. The statistical contents include date, longitude, latitude, and daily yield. The spatial resolution is 1° × 1°.

The climate index PDO was obtained from the website of the Joint Institute for Atmospheric and Oceanic Research (JISAO) at the University of Washington (http://research.jisao.washington.edu/pdo/PDO.latest). The environmental data, including sea surface temperature (SST) and chlorophyll concentration (Chl *a*), were obtained from the Oceanwatch website of National Oceanic and Atmospheric Administration (NOAA) (http://oceanwatch.pifsc.noaa.gov/erddap/index.html). The time range was from January to December of 1998–2016. The data of the spawning field of *Ommastrephes bartramii* are from January to May, and the range is 20°–30°N and 130°–170°E; the data of feeding ground are 35°–50°N and 150°–175°E from July to November. The temporal resolution is monthly, and the spatial resolution is 1° × 1°. The average SST and the average Chl *a* of the spawning grounds and feeding grounds in each month were calculated using the averaging method.

The GM (1, 1) model was established for the CPUE series of different time lengths, and the average relative error of the model established by the CPUE series of each year was calculated. The CPUE series with relatively small errors and variances were selected as the mother series for subsequent modeling.

The environmental and climatic factors during the spawning and feeding periods were analyzed using the gray correlation method. Using the CPUE of the current year as the mother sequence, the SST (abbreviated as SGSST and FGSST, respectively) and Chl *a* concentration (abbreviated as SGC and FGC, respectively) in the spawning ground and the feeding ground and the Pacific interdecadal oscillation index (PDO) were used as the subsequences. The correlation between the mother sequence and each subsequence was calculated, and the one with the largest gray correlation among the monthly indicators was used as a factor in the abundance index prediction model. The calculation method of gray correlation is shown in Chap. 3, and the resolution coefficient is set to 0.5.

The abundance index of *Ommastrephes bartramii* in the northwestern Pacific Ocean was predicted using the discrete GM (0, N) and GM (1, N) models. The numbers 0 and 1 represent the order of the model, and $N = i + 1$ (i is the number of factors). The specific calculation method of the model is shown in Chap. 5. The following six models were designed:

Model 1: GM (0, 6) model that includes all factors, including SGSST, FGSST, SGC, FGC, and PDO;
Model 2: GM (0, 5) model without SGSST;
Model 3: GM (0, 5) model without FGSST;
Model 4: GM (0, 5) model without SGC;
Model 5: GM (0, 5) model without FGC;
Model 6: GM (0, 5) model without PDO;
Model 7: GM (1, 6) model that includes all factors, including SGSST, FGSST, SGC, FGC, and PDO;
Model 8: GM (1, 5) model without SGSST;
Model 9: GM (1, 5) model without FGSST;
Model 10: GM (0, 5) model without SGC;
Model 11: GM (1, 5) model without FGC;
Model 12: GM (1, 5) model without PDO.

The average relative error between the predicted value and the actual value was calculated by comparing the model-fitted CPUE with the actual CPUE value. The data of the last year of the sample were used for model validation. The optimal model was selected based on the fitting accuracy and prediction accuracy of the model.

6.4.4.2.2 Selection of Model Time Series

According to Fig. 6.4, as the time length of the CPUE sequence increases, the average relative error of the GM (1, 1) model basically exhibits an increasing trend, and the variance gradually decreases. The average relative error of the GM (1, 1) model of the 8-year CPUE series is the smallest (6.28%), so the series with the smallest relative error in the 8-year CPUE series (1998–2005) is selected as the mother sequence of the model establishing GM (0, N) and GM. (1, N) to improve the accuracy of model prediction by adding environmental factors.

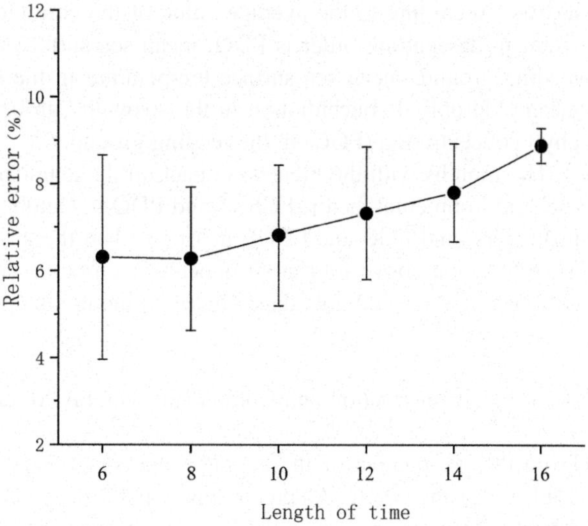

Fig. 6.4 Relative error of the GM (1, 1) prediction model of the CPUE sequence with different time lengths (Xie and Chen 2021b)

Table 6.20 Gray correlation coefficients between the subsequences of each environmental factor and the mother sequence of the current year's CPUE (Xie and Chen 2021b)

Month	Mean sea surface temperature in the spawning ground (SGSST)	Average sea surface temperature in the feeding ground (FGSST)	Average chlorophyll concentration in the spawning ground (SGC)	Average chlorophyll concentration in the feeding ground (FGC)	Pacific Interdecadal Oscillation Index (PDO)
Jan.	0.754	-	0.751	-	0.946
Feb.	0.755	-	0.613	-	0.956
Mar.	0.754	-	0.708	-	0.965
Apr.	0.747	-	0.747	-	0.965
May	0.740	-	0.694	-	0.942
Jun.	-	-	-	-	0.909
Jul.	-	0.727	-	0.651	0.559
Aug.	-	0.767	-	0.694	0.837
Sep.	-	0.732	-	0.620	0.964
Oct.	-	0.794	-	0.677	0.968
Nov.	-	0.786	-	0.643	0.917
Dec.	-	-	-	-	0.898
Average value	0.750	0.761	0.702	0.657	0.902

6.4.4.2.3 Selection of Influencing Factors

According to the results of gray correlation analysis (Table 6.20), the effect of the Pacific Interdecadal Oscillation Index (PDO) on CPUE is the largest, and its average degree of gray correlation is much greater than that of the other four environmental factors. According to the average value of the correlation, the importance of each factor in descending order is PDO, mean sea surface temperature (FGSST) in the feeding ground, mean sea surface temperature in the spawning ground (SGSST), mean chlorophyll concentration in the spawning ground (SGC), and mean chlorophyll concentration (FGC) in the feeding ground.

The months with the greatest impact on the abundance index were different for each environmental factor: FGSST and PDO in October, SGSST in February, SGC in January, and FGC and CPUE in August had the largest gray correlation degree. Therefore, the above four environmental factors are considered the key factors in establishing the squid abundance index of the prediction model.

6.4.4.2.4 Construction and Comparison of Gray Models

From the perspective of the average relative error of the model (Table 6.21 and Table 6.22), the GM (0, *N*) prediction model is higher than the GM (1, *N*) prediction model. The average error in descending order is as follows: (1) GM (0, *N*) model:

Table 6.21 The relative errors of the GM (0, *N*) prediction model for the abundance index of *Ommastrephes bartramii* in the North Pacific Ocean (Xie and Chen 2021b)

Year	Model 1	Model 2	Model 3	Model 4	Model 5	Model 6
1999	3.89	3.78	3.93	3.74	3.16	8.39
2000	5.39	5.22	5.31	3.37	7.69	5.73
2001	0.38	0.01	0.25	3.24	3.42	8.85
2002	3.90	3.64	5.02	8.76	2.07	0.72
2003	0.95	0.76	2.97	3.21	2.75	8.46
2004	1.66	3.39	0.05	2.29	1.29	2.85
2005	1.37	1.78	1.01	2.49	1.09	0.18
Average relative error	2.51	2.65	2.65	3.87	3.07	5.26
Validation	9.00	7.89	9.23	1.18	18.80	7.02

Table 6.22 Relative errors of the GM (1, *N*) prediction model for the abundance index of *Ommastrephes bartramii* (Xie and Chen 2021b)

Year	Model 1	Model 2	Model 3	Model 4	Model 5	Model 6
1999	19.34	16.14	19.10	10.32	28.23	0.37
2000	4.09	0.83	4.14	2.70	4.11	18.15
2001	5.51	3.81	5.65	3.35	10.71	0.78
2002	11.14	12.91	11.01	14.40	14.13	62.42
2003	9.08	7.10	9.45	8.80	10.07	9.02
2004	0.16	4.46	0.70	0.97	4.15	3.89
2005	0.07	4.81	0.44	4.53	4.16	33.22
Average relative error	7.06	7.15	7.21	6.44	10.79	18.27
Validation	28.39	16.42	28.46	1.20	45.79	138.54

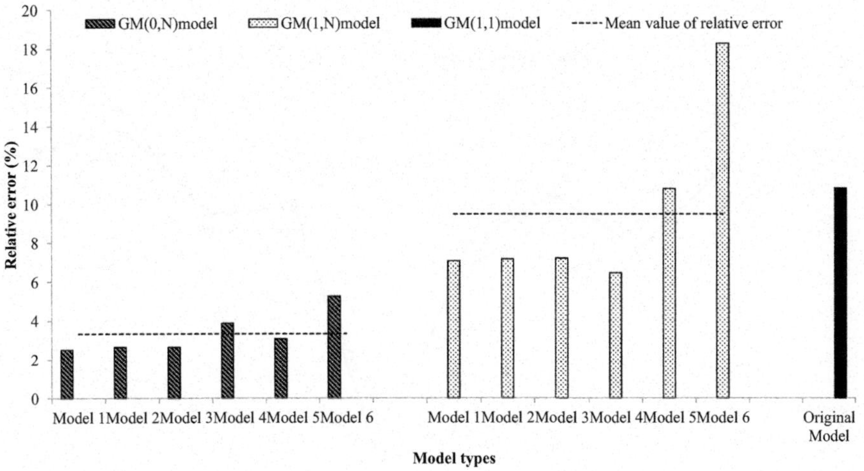

Fig. 6.5 The average relative error of all model types for different orders of prediction models (Xie and Chen 2021b)

Table 6.23 Parameter values of the four GM (0, *N*) models (Xie and Chen 2021b)

	a	SGSST	FGSST	FGC	PDO
Parameters	1.71	0.05	0.30	−51.13	−0.19

model 1 > model 3 > model 2 > model 5 > model 4 > model 6; (2) GM (1, *N*) model: model 4 > model 1 > Model 2 > Model 3 > Model 5 > Model 6 (Fig. 6.5).

Based on the results of model validation in 2006 (Table 6.23), whether it is the GM (0, *N*) model or the GM (1, *N*) model, the prediction accuracy of Model 4 is much higher than that of the other models, with a relative error of 1.18%. The relative errors of the other GM (0, *N*) models are as follows: model 6 (relative error 7.02%), model 2 (relative error 7.89%), model 1 (relative error 9.00%), and model 3 (relative error of 9.00%). The relative errors of the other GM (1, *N*) models are as follows: model 2 (relative error 16.42%), model 1 (relative error 28.39%), model 3 (relative error 28.46%), model 5 (relative error of 45.79%), and model 6 (relative error of 138.54%). Because the fitting error of each model in the GM (0, *N*) model is not large and the validation result of model 4 is much smaller than that of the other models, model 4 without the SGC factor is selected as the best model for predicting the abundance index of *Ommastrephes bartramii* (Fig. 6.6).

In this study, based on gray system theory and methods, the environmental and climatic factors of spawning grounds and feeding grounds were used as indicators to predict the abundance index of *Ommastrephes bartramii*. From the results of the model (Fig. 6.5), the fitting accuracy of almost all models (except for GM (1, *N*) model 6) was greater than that of the GM (1, 1) model, and model 4 (GM (0, *N*)) did not contain the SGC factor and had the best prediction effect. From the perspective of the relationship between the fitted value and the actual value (Fig. 6.6), the

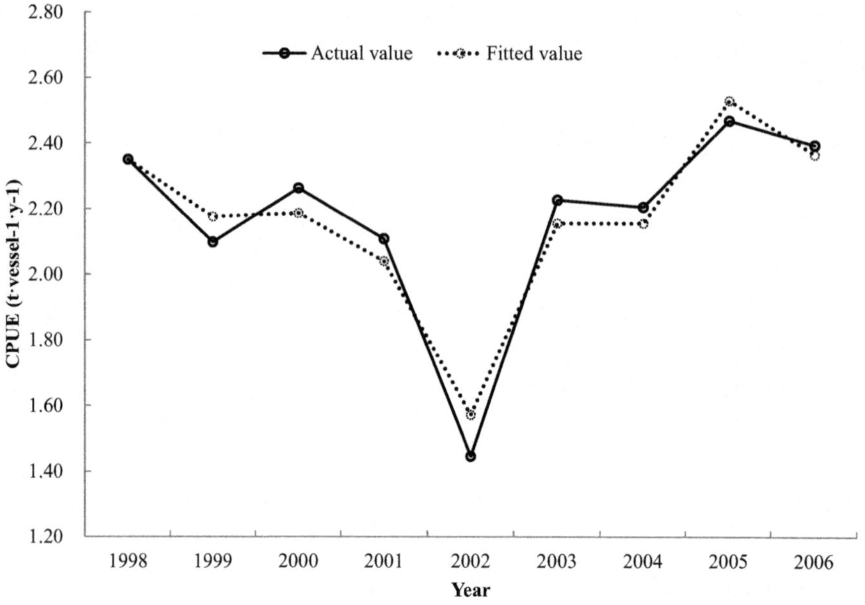

Fig. 6.6 Comparison of the predicted values and the true values of the abundance index of *Ommastrephes bartramii* based on the GM (0, N) model in the North Pacific Ocean (Xie and Chen 2021b)

variation trend of CPUE is basically the same, and the variation amplitude of the fitted value predicted by the model is small. The value of -1.71 (Table 6.23) satisfies the conditions of the medium- and long-term forecast model ($-a < 0.3$), indicating that the abundance index of *Ommastrephes bartramii* in the northern Pacific Ocean is indeed affected by marine climate factors and environmental factors.

In this study, the preselection of the early-stage data of the prediction model and the selection of the later-stage model were optimized, and good results were obtained. The gray system model has the advantage of allowing a small sample size and does not require a priori information. However, it can be seen from the results (Fig. 6.4) that the selection of the sample size has a certain range of application, and a sample size that is too small or too large will affect it. The accuracy of the final prediction model. In addition, the prediction results of GM models with different orders are somewhat different, and the results show that the prediction results of the 0-order GM (0, N) model are better than those of the first-order GM (1, N) model (Fig. 6.5). This is not universal, and the fitting accuracy results for CPUE sequences with different characteristics may be different. In summary, in the construction of the gray prediction model, selecting the appropriate original data series, identifying the key affecting factors, and comparing and screening a variety of different types of models can more accurately predict the changes in the abundance index of *Ommastrephes bartramii* in the northern Pacific Ocean, which will provide technical support for fishery production.

6.4.4.3 Resource Forecasting of the Pacific Stock of *Scomber australasicus* Based on the Gray System

Australian mackerel, *Scomber australasicus*, is widely found in the Pacific Ocean, and the major fishing countries include Japan, South Korea, and China. As an important economic species in various countries, it is particularly important to grasp the changes in the amount of Australian mackerel resources and to scientifically predict the amount of resources in fishery production and scientific management. To this end, Zhang and Chen (2019) used the gray correlation and gray prediction model to analyze the effect of environmental factors on the resources of Australian mackerel based on the resource assessment data provided by the Central Fisheries Research Institute of Japan. The gray forecast of Australian mackerel resources can provide technical support for the sustainable development and scientific management of Australian mackerel.

6.4.4.3.1 Materials and Methods

The resource data of Australian mackerel are from the Resource assessment report of the Australian Mackerel in 2015. The time period is from 1995 to 2014, and the data are the actual resource and catch data of Australian mackerel in this study. The resource data of 1995–2012 were used for modeling, and the resource data of 2013 and 2014 were used for verification and comparison.

The marine environmental data include surface temperature (SST), Kuroshio tidal level difference, and Pacific interdecadal oscillation (PDO). In this study, according to the distribution area in the resource assessment report of Australian mackerel, the area (140°E–160°E, 35°N–50°N) was used as the feeding ground. Two fields (130°E–132°E, 30°N–32°N and 138°E-141°E, 34°N–35°N) were selected as spawning grounds. In this study, SPSS19 software was used to analyze the correlation between the monthly average temperature and the amount of resources in each region. The monthly temperature with the highest linear correlation coefficient was selected as the temperature (SST1) on the feeding ground and the temperature (SST2, SST3) on the spawning ground. The spatial resolution of the surface temperature data is $1° \times 1°$, and the temporal resolution is monthly. The data are from the website http://iridl.ldeo.columbia.edu/.

Studies have shown that the Kuroshio has a significant impact on pelagic fishery resources. The strength of the Kuroshio Current is expressed by the tidal level difference of the Kuroshio Current, and the annual average tidal range data are selected from the website. The data are from the website http://www.data.jma.go.jp/.

The Pacific interdecadal oscillation PDO (PDO) is a climate change mode on an interdecadal time scale. The PDO can directly cause interdecadal variability in the climate in the Pacific Ocean and its surrounding areas and has an important modulating effect on interannual variabilities, such as El Niño-Southern Oscillation

(ENSO). Therefore, the annual average PDO is selected in this study. The data are from the website http://www.research.jisao.washington.edu/pdo/PDO.latest.

The data were averaged and then subjected to general correlation analysis. See Chap. 3 for the calculation method. The resolution coefficient is 0.1. In this study, one GM (1, 1) model that does not consider environmental factors and four GM (1, 2) models that consider environmental factors are established, which are the GM (1, 2) model based on SST1, the GM (1, 2) based on SST2, the GM (1, 2) model based on SST3, the GM (1, 2) model based on the tidal level difference, and the GM (1, 5) model based on SST1, SST2, SST3 and the tidal level difference. The modeling process of GM (1, 1) and GM (1, N) can be found in Chap. 5.

6.4.4.3.2 Correlation Analysis of Monthly Temperature Factors and Resources of Australian Mackerel

The correlation between the SSTs of the spawning ground and the feeding ground and the resources of Australian mackerel in each month (Table 6.24) shows that the correlation between the SST and the amount of resources in the feeding ground in August was the highest at 0.42, and the correlation between SST2 in Jan, SST3 in May and the resource amount of Australian mackerel were the highest at 0.6 and 0.52, respectively. According to the results of previous studies, Australian mackerel spawns in winter and spring and enters the feeding grounds from late June to early September. Therefore, SST1 in August was selected to characterize the temperature characteristics of the feeding ground, and SST2 in Jan and SST3 in May on the spawning ground represent the temperature characteristics on the spawning ground.

6.4.4.3.3 Analysis of Environmental Factors Affecting the Resources of Australian Mackerel Based on Gray Correlation

The averaging transformation is performed on each original series, and the gray correlation degree is calculated. The gray correlation degree of each factor can be obtained as follows:

$$L(\text{SST1}) = 0.8791$$

$$L(\text{SST2}) = 0.8709$$

$$L(\text{SST3}) = 0.8703$$

$$L(\text{tidal level difference}) = 0.8597$$

$$L(\text{PDO}) = 0.2312$$

The analysis shows that L(SST1) > L(SST2) > L(SST3) > L(Tide level difference) > L(PDO). If $L > 0.6$ is selected as the environmental factor for the

Table 6.24 The correlation coefficient between monthly SST and Australian mackerel resources (Zhang and Chen 2019)

Month	Jan.	Feb.	March	April	May	June	July	Aug.	Sep.	Oct.	Nov.	Dec.
SST1	0.19	0.34	0.01	0.17	0.21	0.22	0.21	0.42	0.28	0.09	0.16	0.19
SST2	0.60	0.26	0.01	0.26	0.48	0.23	0.02	0.11	0.07	0.02	0.28	0.58
SST3	0.50	0.37	0.26	0.24	0.52	0.43	0.07	0.08	0.18	0.30	0.15	0.27

establishment of the GM model, SST1, SST2, SST3, and the tide level difference are used as environmental factors for establishing the model.

6.4.4.3.4 Establishment of a Gray Forecast Model for Australian Mackerel Resources

Model 1: The GM (1, 1) model that does not consider environmental factors. The GM (1, 1) model calculation was performed using the resource amount data, the gray parameters $a = -0.01280$, $b = 250.2298$, and the response function $B(t + 1) = 19861.4365\exp(0.01280\ t) - 19544.7203$, with an average error of 18.65%.

Model 2: The GM (1, 2) model based on the temperature of the feeding field SST1. The calculation shows that the gray parameters $a = 0.09926$ and $b1 = 4.6495$, and the response function is $B(t + 1) = (309.0000 - 46.8400 * \mathrm{SST1})\mathrm{EXP}(-0.09926\ t) + 46.8400 * \mathrm{SST1}$, with an average error of 28.53%.

Model 3: The GM (1, 2) model based on the spawning field temperature SST2. The calculation shows that the gray parameters $a = 0.08862$ and $b1 = 3.9551$, and the response function is $B(t + 1) = (309.0000 - 44.6291 * \mathrm{SST2})\mathrm{EXP}(-0.08862\ t) + 44.6291 * \mathrm{SST2}$, with an average error of 28.93%.

Model 4: GM (1, 2) model based on spawning field temperature SST3. The calculation shows that the gray parameters $a = 0.08999$ and $b1 = 3.9453$, and the response function is $\mathrm{B}(t + 1) = (309.0000 - 43.8408 * \mathrm{SST3})\mathrm{EXP}(-0.08999\ t) + 43.8408 * \mathrm{SST3}$, with an average error of 29.46%.

Model 5: The GM (1, 2) model based on the tidal level difference (T). The calculation shows that the gray parameters are $a = 0.1269$ and $b1 = 7.0036$, and the response function is $B(t + 1) = (309.0000 - 55.2094 * T)\mathrm{EXP}(-0.1269\ t) + 55.2094 * T$, with an average error of 33.79%.

Model 6: The GM (1, 5) model based on the feeding ground SST1, the spawning ground SST2, the spawning ground SST3, and tidal level difference (T). The calculation shows that the gray parameters $a = 0.2618$, $b_1 = 58.7562$, $b_2 = -47.6616$, $b_3 = -3.4524$, and $b_4 = 8.7473$, and the response function is $B(t + 1) = (309.0000 - 224.4268 * \mathrm{SST1} + 182.04945) * \mathrm{SST2} + 13.18674 * \mathrm{SST3} - 33.41153 * T)\ \mathrm{EXP}(-0.26181\ t) + 224.42684 * \mathrm{SST1} - 182.04945 * \mathrm{SST2} - 13.18674 * \mathrm{SST3} + 33.41153 * T$, with an average error of 33.79%.

6.4.4.3.5 Comparison and Validation of Gray Prediction Models

The above model was used to evaluate the amount of resources of Australian mackerel in 2013 and 2014. The specific results are shown in Table 6.25. The average prediction error of the GM (1, 2) model based on SST1 is the smallest, which is only 3.73%. The average prediction error of the GM (1, 2) model based on SST2 is 4.41%. The average prediction error of the GM (1, 2) model based on SST2

Table 6.25 Prediction of Australian mackerel resources in 2013 and 2014 (Zhang and Chen 2019) Unit: thousand tonnes

	Model 1	Model 2	Model 3	Model 4	Model 5	Model 6
2013	838.95	810.26	768.48	784.87	498.39	792.62
2014	896.34	822.01	788.65	767.36	936.88	1052.95
Mean error (%)	6.72	3.73	4.41	4.78	29.56	19.38

was 4.78%. The average prediction error of the GM (1, 2) model based on the tidal level difference was 29.56%.

The results show that the gray prediction model established based on the SST of the feeding ground and spawning ground has a high accuracy in forecasting the resources of Australian mackerel and can be applied to subsequent fishery production. Analysis of the gray parameter values a and b of the GM (1, 5) model shows that among all the factors, SST2 and SST3 have the highest impact on the amount of resources of Australian mackerel.

6.4.4.4 Gray Prediction of Diseases of Large Yellow Croaker in Cage Culture

Prevention of farmed animal diseases is one of the keys to the healthy and sustainable development of the aquaculture industry. Owen et al. (2013) introduced the gray system theory to explore the occurrence and development of bacterial diseases in cage-cultured large yellow croaker and their relationship with environmental factors and established a gray model for the prediction of bacterial diseases in cage-cultured large yellow croaker. It is expected to provide a method for the prediction of bacterial diseases in cage cultures of large yellow croaker and provide ideas and approaches for the disease prediction of other aquaculture organisms.

Table 6.26 shows the incidence of bacterial diseases in cage-cultured large yellow croaker in Zhoushan, Zhejiang Province, China, from 2001 to 2008. Analysis of the incidence of bacterial diseases in large yellow croaker from 2001 to 2006 showed that there was a large-scale outbreak of disease every year. The diseased water body in the whole city was above 3000 m^3, and the highest incidence was 36,000 m^3. If we can predict in advance and take active preventive measures, it is possible to reduce the scope of the disease or avoid the occurrence of the disease.

From 2001 to 2008, we monitored the aquaculture, morbidity, and mortality of large yellow croaker in cage culture in Dinghai District, Daishan District, and Putuo District of Zhoushan City. At the same time, sampling points were set up in these three areas, and the physical and chemical factors and biological factors at these sampling points were regularly measured, including water temperature, salinity, suspended matter, dissolved oxygen, pH, phosphate, silicate, nitrogen, ammonia nitrogen, inorganic nitrogen, chemical oxygen demand (COD), zooplankton, and phytoplankton species and quantity. The specific calculation of gray correlation is

Table 6.26 The incidence of bacterial diseases in large yellow croaker in 2001–2008 (Owen et al. 2013)

Year	Month											
	Jan.	Feb.	March	April	May	June	July	Aug.	Sep.	Oct.	Nov.	Dec.
2001	0	0	0	0	0	0	0	0	4002	0	0	0
2002	0	0	0	0	0	372	1939	1656	3386	1322	0	0
2003	0	0	0	0	210	270	1500	5300	2400	672	0	0
2004	0	400	0	1500	0	0	0	2700	1200	30,000	1500	0
2005	0	0	0	0	0	0	8640	0	0	0	0	0
2006	0	0	0	0	0	0	0	36,000	18,000	2400	2400	
2007	0	0	0	0	0	0	0	0	0	0	0	0
2008	0	0	0	0	0	0	2000	0	0	0	0	0

shown in Chap. 2, and its resolution coefficient is set to 0.5. See Chap. 5 for the GM (1, N) model.

6.4.4.4.1 Correlation Analysis of the Morbidity of Large Yellow Croaker and Environmental Factors

To make full use of the information provided by changes in environmental factors on the occurrence and development of bacterial diseases in large yellow croaker, it is first necessary to understand the relationship between environmental factors and their changes and the occurrence and development of bacterial diseases in large yellow croaker. The incidence sequence of disease X_0 is related to water temperature, salinity, suspended matter, dissolved oxygen, pH, phosphate, silicate, nitrate nitrogen, nitrite nitrogen, ammonia nitrogen, inorganic nitrogen, COD, zooplankton, phytoplankton, etc. The degree of correlation γ (X_0, X_i) of the environmental factors is shown in the first column of Table 6.27. Table 6.27 shows that the occurrence of diseases in large yellow croaker was correlated with changes in 14 environmental factors to varying degrees. The correlation degree above 0.8 is water temperature, suspended matter, COD, inorganic nitrogen, nitrate nitrogen, nitrite nitrogen, and ammonia nitrogen. Changes in these environmental factors are associated with the occurrence and development of bacterial diseases in large yellow croaker. However, these changing trends are not necessarily synchronized. When the external environment changes, such as the climate, some environmental factors in aquaculture waters change first. When these changes reach a certain level and exceed the physiological limit of fish, they may cause disease. To effectively monitor and warn of the occurrence of fish diseases, it is necessary to consider the advanced relationship between the changes in environmental factors and the occurrence of fish diseases,

Table 6.27 Correlation between the incidence of bacterial diseases and environmental factors (Owen et al. 2013)

Factor	Correlation degree	
	$\gamma(X_0, X_i)$	$\gamma'(X_0, X_i)$
Water temperature/°C	0.867	0.9067
Salinity	0.6657	0.767
Suspended matter/(mg/L)	0.8564	0.8684
Dissolved oxygen/(mg/L)	0.7734	0.7709
pH	0.7953	0.7817
Phosphate/(mg/L)	0.799	0.7655
Silicate/(mg/L)	0.7214	0.7704
Nitrate/(mg/L)	0.8093	0.7201
Nitrite/(mg/L)	0.8075	0.7389
Ammonia/(mg/L)	0.8027	0.7209
Inorganic nitrogen/(mg/L)	0.8114	0.9037
COD/(mg/L)	0.844	0.8372
Phytoplankton/L	0.6895	0.6758
Zooplankton/(mg/m^3)	0.7034	0.7045

which is the leading indicator of fish diseases. Therefore, the degree of association between the environmental factor sequence and the morbidity sequence is calculated in advance by one order and is denoted as γ' (X_0, X_i). The results obtained are shown in the second column of Table 6.27. Table 6.27 shows that water temperature, suspended matter, inorganic nitrogen, and COD are the factors with γ' (X_0, X_i) greater than 0.8. Therefore, these four factors are used as predictors of bacterial diseases in large yellow croaker.

6.4.4.4.2 Establishment of the GM (1, *N*) Model

Considering that there are many reasons for the occurrence of fish diseases, in addition to their own problems, such as fish constitution, the changes in the environmental factors of aquaculture waters are the incentives for the occurrence of diseases. The establishment of the GM (1, *N*) model considers the establishment of the dynamic relationship between the incidence of large yellow croaker and environmental factors. The GM (1, *N*) model was constructed based on the characteristic data series of the incidence rate from May to October 2003, and the series of water temperature, suspended matter, inorganic nitrogen, and COD were used to establish the GM (1, 5) model. The fitted values of the model were calculated, and the obtained simulation series are listed in columns 1–3 of Tables 6.28 together with the primary incidence series and the residuals of each observation point. The average relative error of the calculation model was 7.9791%. The average accuracy of the forecast is 92.0209%.

In general, when the correlation factor series with a high degree of correlation with the feature data series is introduced into the model, it will provide more information for the prediction of the development trend of the feature data series. However, the increase in the series of relevant factors will also increase the risk of forecasting, especially when the correlation degree of these series of relevant factors is very large. Due to the fluctuation of these series, the volatility of the model may be increased, and the forecast error will increase. At the same time, the increase in the series of relevant factors in the model will increase the difficulty of application.

Table 6.28 Predicted values and fitting residuals of GM (1, 5), GM (1, 4), and GM (1, 3) (Owen et al. 2013)

	GM (1, 5)		GM (1, 4)		GM (1, 3)	
Observed value	Simulation value	Residual	Simulation value	Residual	Simulation value	Residual
10.05	10.0500		10.0500		10.05	
1.16	0.9821	0.1779	0.8748	0.2852	1.2046	−0.0446
6.03	5.8205	0.2095	6.0831	−0.0531	5.6076	0.4224
22.08	21.3068	0.7732	22.1177	−0.0377	21.337	0.734
10.00	10.6457	−0.6457	9.8093	0.1907	10.7148	−0.7148
3.17	3.5227	−0.3527	3.7466	−0.5766	3.3335	−0.1635

Therefore, a good model should ensure the fit of the model while keeping the forecast variables as small as possible. Therefore, three and two different combinations of these four correlation factors were selected to establish a GM (1, N) ($N = 3$, 4) model, and four GM (1, 4) models and six GMs were obtained. (1, 3) model. The relative errors of each group were compared. The average relative error of the four GM (1, 4) models was 8.0262–11.1136%, which was higher than that of the GM (1, 5) model. The model with the smallest average relative error contains suspended matter, inorganic nitrogen, and COD. Comparing the six GM (1, 3) models, the average relative error of the five models was higher than that of the GM (1, 5) model, which was 11.707–62.8392%. The model is composed of inorganic nitrogen and COD, and the expression of its G (1, 3) model is:

$$\begin{cases} \widehat{x}^{(1)}(k+1)=\Big\{10.05-\frac{1}{2.2092}\Big[-54.7598x_1^{(1)}(k+1)+56.4861x_2^{(1)}(k+1)\Big]e^{-2.3307k} \\ \qquad +\frac{1}{2.2092}\Big[-54.7598x_1^{(1)}(k+1)+56.4861x_2^{(1)}(k+1)\Big]\Big\} \\ \widehat{x}^{(0)}(k+1)=\widehat{x}^{(1)}(k+1)-\widehat{x}^{(1)}(k) \end{cases}$$

where $X_1{}^{(0)}$ and $X_2{}^{(0)}$ are the inorganic nitrogen and COD sequences, respectively. The simulation values and residuals of the optimal model are shown in columns 4–7 of Tables 6.28.

According to the established GM (1, N) model, as long as the values of the relevant factors in the current period are measured, the incidence of bacterial diseases in cage-cultured large yellow croaker in the next period can be predicted. Comparing the GM (1, 5) model and the GM (1, 3) model, the GM (l, 3) model lacks the two environmental factor sequences of water temperature and suspended matter, but the fitted residuals of GM (1, 3) are better than those of GM (1, 5). As a leading indicator of water temperature, the correlation between water temperature and the occurrence of bacterial diseases in large yellow croaker was the largest. The increase in water temperature is a prerequisite for the occurrence of bacterial diseases in large yellow croaker. When the water temperature reaches a certain range, the occurrence and development of the disease depends on the water quality of the aquaculture waters. The increase in the amount of suspended matter will affect the photosynthesis of phytoplankton, increase the consumption of organic matter, and change the physical and chemical properties of the water body, such as inorganic nitrogen and COD. Its effect on the incidence is mainly reflected in the changes in inorganic nitrogen and COD. Therefore, two environmental factors, water temperature and suspended matter, were added to the model, which, in contrast, increased the uncertainty of the model and reduced the fitting accuracy.

6.4.4.5 Prediction of Gray Catastrophe of Fishery Resources

6.4.4.5.1 Prediction of Gray Catastrophe for the Resource Abundance of *Ommastrephes bartramii*

As an important economic cephalopod in the northwestern Pacific Ocean, the resources of *Ommastrephes bartramii* have drastically changed over the years and are difficult to predict. Therefore, Xie and Chen (2020) attempted to establish a catastrophe prediction model for the abundance of *Ommastrephes bartramii* to provide a basis for national exploitation based on the gray theory.

Data and Methods

The fishery production data were from the Chinese Squid Fishing vessels. The time period was from June to November of 1995 to 2017. The spatial range was 35°–45° N and 140°–179°E. The statistical contents included date, longitude, and latitude. The spatial resolution is 1° × 1°. The catch per unit effort (CPUE) was used to characterize the abundance index of *Ommastrephes bartramii*. With 1° × 1° as a fishing area, the CPUE value of each fishing area in each year was calculated.

The total catch obtained in each year was counted, and the number of fishing vessels in each year was used as the fishing effort. The generalized linear regression model (GLM) was used to normalize the nominal CPUE. The expression of GLM is glm[CPUE] ~ SST + Lon + Lat + ε.

The GM (1, 1) model was used to predict the abundance index of *Ommastrephes bartramii* in the North Pacific. The specific method is as follows:

Set the original sequence $X = (x(1), x(2) ..., x(n))$, and take the third quantile Q_3 of the sequence (a value greater than 75%) as the upper limit catastrophe value ξ_1, and the first quantile Q_1 of the series (a value less than 25%) is taken as the lower catastrophe value $\xi_{2.}$ The points greater or less than ξ in sequence X are considered outliers, and they are composed of the upper limit or the lower limit catastrophe sequence $X_\xi = (x[q(1)], x[q(2)]..., x[q(m)])$. Let the catastrophe date series $Q^{(0)} = (Q(1), q(2)..., q(m))$, and a GM (1, 1) model is established for disaster prediction. The GM (1, 1) model establishment method is described in Chap. 5.

CPUE Distribution and Its Catastrophe Point

The nominal average CPUE and normalized CPUE for each year from 1995 to 2017 are shown in Fig. 6.7. The normalized CPUE of the GLM showed basically the same variation trend as the nominal CPUE, but the fluctuation was small. In 2007, the nominal CPUE value was the largest, which was 4.302 t. d $^{-1,}$ and the nominal CPUE in 2009 was the smallest, which was 1.307 t. d $^{-1.}$ The standardized CPUE value was the largest in 2017, which was 2.915 t. d $^{-1,}$ and the nominal CPUE was the smallest in 2002, which was 1.399 t. d $^{-1}$. According to the classification of catastrophe values, the upper catastrophe points are 2005, 2007, 2008, 2010, 2014,

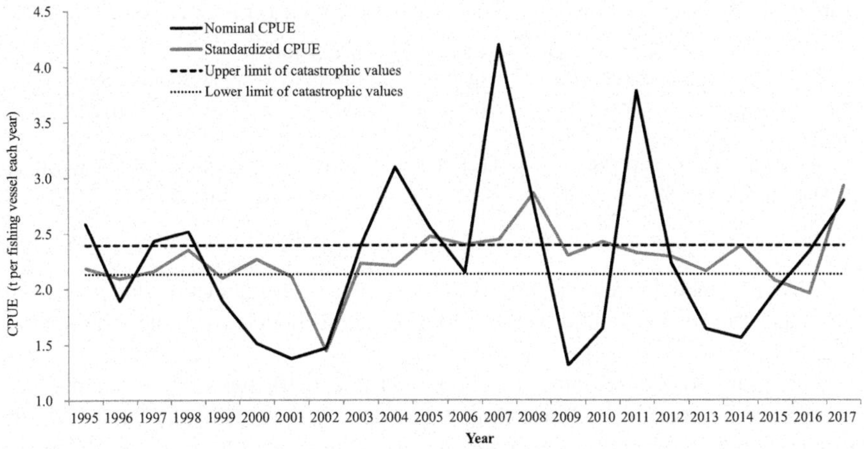

Fig. 6.7 Distribution of the resource abundance of *Ommastrephes bartramii* in the Northwest Pacific Ocean from 1995 to 2017 (Xie and Chen 2020)

Table 6.29 The relative error between the predicted value and the actual value of *Ommastrephes bartramii* (Xie and Chen 2020)

Lower limit catastrophe sequence number	Fitted value	Relative error	Upper limit catastrophe serial number	Fitted value	Relative error
X_5	5.08	1.67	X_{13}	12.26	5.68
X_7	7.47	6.73	X_{14}	14.32	2.28
X_8	10.98	37.25	X_{16}	16.72	4.52
X_{21}	16.14	23.16	X_{20}	19.53	2.35
X_{22}	23.72	7.80	X_{23}	22.81	0.86
Average relative error		15.32			3.13

and 2017, and the lower catastrophe points are 1996, 1999, 2001, 2002, 2015, and 2016.

Establishment and Validation of the Gray Catastrophe Model

The GM (1, 1) model is constructed using the upper catastrophe point and the lower catastrophe point. From the perspective of model accuracy (Table 6.29), the average correlation error of the lower catastrophe prediction model is 15.32%, and the average correlation error of the upper catastrophe prediction model is only 3.13%. From the perspective of the relevant parameters of the model, the development coefficient a reflects the degree of coordination between the main factor and each subfactor. When $a \leq 0.3$, the model can be used for medium- and long-term prediction; when $0.3 < -a \leq 0.5$. The lower-bound catastrophe prediction model ($a = -0.385$) can be used for short-term prediction, and the lower-bound catastrophe prediction model ($a = -0.155$) can be used for medium- and long-term prediction (Table 6.29). When the posterior ratio $c < 0.35$ and the small error probability

Table 6.30 Relevant parameters of the gray catastrophe GM (1, 1) model (Xie and Chen 2020)

	Development coefficient *a*	Posterior ratio c	Small error probability *p*	Response function
Lower limit catastrophe model	−0.385	0.344	1.000	$X_{(t+1)} = 10.823\exp(0.385\ t) - 8.823$
Upper limit catastrophe model	−0.155	0.128	1.000	$X_{(t+1)} = 73.058\exp(0.155\ t) - 62.058$

$p > 0.95$, the model is reliable, the accuracy level is level 1, and the c and p values of the upper- and lower-bound catastrophe prediction models meet the requirements (Table 6.29).

According to the response function in Table 6.30, the time when the next occurrence number of the lower limit catastrophe is approximately 34.86, i.e., the resource abundance under year will occur approximately 12 years after the occurrence; the next occurrence number of the upper limit catastrophe is approximately 26.64. The resource abundance year will occur in the fourth year after the occurrence of (Table 6.30). From the analysis of the average error of the model (Table 6.30), the GM (1, 1) model can effectively predict the occurrence time of a catastrophe.

6.4.4.5.2 Gray Catastrophe Prediction of the Abundance of *Illex argentinus* in the Waters of the Malvinas Islands

Argentine flying squid (*Illex argentinus*) are shallow oceanic species and important economic species. It is particularly abundant at 35°–52°S. It is currently one of the most important cephalopod resources in the world. Among them, the waters of the Malvinas Islands are one of the important fishing grounds of *Illex argentinus*. The average annual catch in this sea area is 200 thousand t, of which squid production accounts for approximately 75% of the total. In high-yield years, the catch around the Malvinas Islands provides 10% of the world's total squid. There is significant interannual variation in the amount of *Illex argentinus*, which may be because the early life history and habitat are highly susceptible to the impact of the marine environment and climatic factors. To this end, Xu et al. (2022) used the gray catastrophe prediction GM (1, 1) model in gray system theory to scientifically predict the year of highest or lowest catch of *Illex argentinus*, which will provide a reasonable scientific basis for the management and sustainable development of fishery resources.

Data and Research Methods

The fishery data in this study are from the annual fishery statistics report of the Fisheries Bureau of the Malvinas Islands. The data are the annual catch and the number of vessels operated from 1995 to 2019. The annual fishing yield per vessel (CPUE, t/ship) is used to characterize the abundance of *Illex argentinus*. The GM

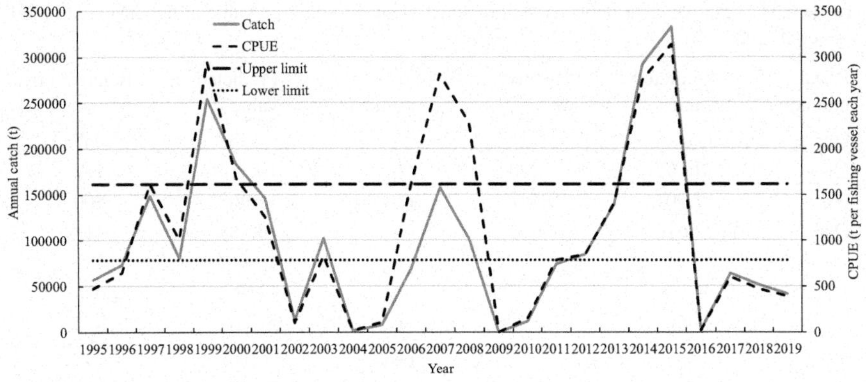

Fig. 6.8 Catch and CPUE of *Illex angentinus* in the Malvinas Islands from 1995 to 2019

(1, 1) model was used to predict the abundance of *Illex argentinus* resources in the Malvinas Islands. The specific modeling process and model verification are shown in Chap. 5.

Changes in the Production and CPUE of Illex argentinus

According to the statistical analysis, during the 25 years from 1995 to 2019, the annual production of *Illex argentinus* in the waters of the Malvinas Islands fluctuated significantly (Fig. 6.8). The annual average production of *Illex argentinus* in 2009 was the lowest, only 3 tons. In the following years, the yield began to rise and reached the highest yield in 2015, reaching 332,863 t. In 2016, the yield decreased sharply to only 2297 t, and the yield in the following years was at a low level.

During the period from 1995 to 2019, the average annual CPUE of *Illex argentinus* was 1118.807 t/ship, and the variation trend of CPUE was basically consistent with the variation trend of production. In 2009, the CPUE value was the smallest, only 0.143 t/ship. In the following years, the CPUE began to rise. In 2015, the CPUE reached a maximum value of 3140.217 t/ship. In 2016, the CPUE decreased sharply to only 22.086 t/ship. However, the CPUE in 2007 and 2008 was abnormal. Although the annual production was low, the CPUE was relatively high. The reason may be related to the decrease in the number of fishing vessels operating in that year.

Establishment and Testing of the GM (1, 1) Model

According to the calculation results of the upper and lower limits of the catastrophic value (the upper catastrophe value is 1615 t/ship, and the lower catastrophe value is 784 t/ship), the years of rich years are determined to be 1999, 2000, 2007, 2008, 2014, and 2015. The corresponding time series are 5, 6, 13, 14, 20, and 21; the years of poor years are 1995, 1996, 2004, 2005, 2009, 2010, 2016, 2017, 2018, and 2019. The corresponding time series are 1, 2, 8, 10, 11, 15, 16, 22, 23, 24, and 25.

Table 6.31 Relative error between the predicted value and the true value of the GM (1, 1) model (Xu et al. 2022)

Serial number	Predicted value	Relative error/%	Serial number	Predicted value	Relative error/%
Q_{13}	11.214	13.741	Q_8	8.887	11.089
Q_{14}	14.168	1.200	Q_{10}	10.313	3.132
Q_{20}	17.901	10.497	Q_{11}	11.968	8.800
Q_{21}	22.616	7.698	Q_{15}	13.888	7.411
			Q_{16}	16.117	0.731
			Q_{22}	18.703	14.986
			Q_{23}	21.704	5.634
			Q_{24}	25.187	4.945
			Q_{25}	29.228	16.913
Average relative error		8.284	Average relative error		8.182

Table 6.32 Related parameters and prediction results of the GM (1, 1) model (Xu et al. 2022)

Model	Development coefficient (a)	Posterior ratio(c)	Small error probability(P)	Catastrophe point	Catastrophe point number
Rich year	−0.234	0.285	1.000	$X_0(7)$	28.57
				$X_0(8)$	36.10
				$X_0(9)$	45.61
Poor year	−0.149	0.288	1.000	$X_0(12)$	33.92
				$X_0(13)$	39.36
				$X_0(14)$	45.677

The GM (1, 1) model for forecasting rich years is $X_{(t)} = 33.689\exp(0.234\ t) - 28.689$.

The GM (1, 1) model for the prediction of poor years is $X_{(t)} = 57.527\exp(0.129\ t) - 56.527$.

The relative error between the predicted value of the catastrophe model and the actual value is shown in Tables 6.31. For the GM (1, 1) model for forecasting rich years, the average relative error was 1.20–13.8%, and the average relative error was 8.824%. For the GM (1, 1) model for forecasting poor years, the average relative error was 3.13–16.92%, and the average relative error was 8.182%. The prediction parameters of the model are shown in Tables 6.32. In the prediction model for the abundance catastrophe of *Illex argentinus* in this study, both the rich year forecasting model ($a = -0.234$) and the poor year forecasting model ($a = -0.149$) can be used for medium- and long-term forecasting. The variance ratio was less than 0.35, the small probability error P value was 1.000, greater than 0.95 (Table 6.32), and the accuracy of the model was grade I.

The last abundance year was 2015, and the corresponding serial numbers for the next three occurrences that exceeded the catastrophic point threshold were 28.57, 36.10, and 45.61 (Table 6.32). Therefore, the years of rich resources in the future are

2021, 2028, and 2038. Similarly, the last apocalyptic year was 2019, and the numbers corresponding to the next three occurrences that exceeded the catastrophic point threshold were 33.92, 39.36, and 45.677 (Table 6.32); therefore, the years with poor resources in the future abundance are 2024, 2029, and 2036.

Through the processing of the original CPUE data of *Illex angentinus* in the waters of the Malvinas Islands from 1995 to 2019 and the use of the gray catastrophe model to better predict the rich years and poor years, the average relative errors of the models for the rich years and the poor years are 8.284%. This result can provide guidance for fishery production. According to relevant data in 2021, the fishing yield of *Illex angentinus* in the waters of the Malvinas Islands reached 172,000 tons, and the average CPUE exceeded the upper limit of the catastrophe value of 1615 t/ship, which is a good year. The predicted value is credible.

There are many factors that affect the establishment of the catastrophe prediction model, such as the delineation of the upper and lower limits of the catastrophe, the length of the prediction time, and the environmental factors. The determination of the upper and lower limits of the catastrophe, as well as the selection and optimization of the prediction time, can be compared and determined by establishing different catastrophe prediction models to overcome this problem. In addition, in the subsequent establishment of catastrophe prediction models, it is possible to consider the climate and environmental factors that affect the abundance of *Illex angentinus*.

References

Chen XJ, Zhou YQ (2001) Analysis and forecast of manpower resources in Chinese Marine Fisheries by using grey theory. J Zhanjiang Ocean Univ 21(1):22–29. (In Chinese)

Guo M (1992) Gray forecast of shrimp yield in Bohai Sea. Fish Sci 3:10–14. (In Chinese)

Liu SF, Guo TB, Dang YG (1999) Gray system theory and its application. Science Press. (In Chinese)

Lu Q, Fang Z, Li N et al (2022) Prediction model of fisheries catch based on GM (1, N) in the Indian Ocean. J Fish China:1–8. (In Chinese)

Owen MK, Ni HE, Wang LG et al (2013) A forecasting model for bacterial disease of cage cultured large yellow croaker (*Pseudosciaena crocea*) based on grey system theory. J Fish China 37(6): 920–926. (In Chinese)

Xie MY, Chen XJ (2020) Grey catastrophe year prediction for the abundance of neon flying squid (*Ommastrephes bartramii*) in the Northwest Pacific. Haiyang Xuebao 42(4):40–46. (In Chinese)

Xie MY, Chen XJ (2021a) Analysis of the fishing seasons characteristics of *Ommastrephes bartramii* and prediction of the main fishing seasons based on the grey system theory. Prog Fish Sci 42(4):1–8. (In Chinese)

Xie MY, Chen XJ (2021b) Prediction of abundance index of *Ommastrephes bartramii* in the North Pacific Ocean based on different order grey system models. J Shanghai Ocean Univ 30(4): 755–762. (In Chinese)

Xu ZA, Xie MY, Chen XJ (2022) Grey catastrophe year prediction for the abundance index of *Illes angentinus* in the waters near Malvinas islands. J Shanghai Ocean Univ 31(3):642–649. (In Chinese)

Zhang C, Chen XJ (2019) Forecasting model for spotted mackerel biomass based on grey system theory. J Shanghai Ocean Univ 28(1):154–160. (In Chinese)

Chapter 7
Gray Decision

Xinjun Chen

Abstract Decision-making is making a choice or a decision, and it is a key step in the process of solving a problem. Decision-making has both its technical side, that is, the search process of alternatives, and decision-making has its adaptive side, that is, the selection process of alternatives. Decision-making is not only an individual behavior, but also an organizational behavior. In the actual management activities, the decision-making is influenced by the organizational environment, to avoid making bad decisions. According to Herbert Symon, the essence of management is decision-making, which runs through the whole process of management and determines the success or failure of the whole management activity. Usually, because of the complexity of the social economic system and the crisscross of decision factors, it is difficult for any decision maker to make the optimal decision based on intuition and experience. Therefore, in the modern scientific decision-making, often with the help of the method of natural science, the use of mathematical tools, the establishment of the relationship between the decision-making variables formula and model, to reflect the essence of the decision-making problem, simplify complex decision problems. The general form of the decision model is $V = F(A_i, S_j)$, in which V is the value goal, A_i is the controllable decision factor, and S_j is the uncontrollable decision factor. In order to obtain the production plan or plan with the greatest benefit, the decision-making model is listed first, and then the optimal one is obtained through the model. In the process of decision-making, many uncertain factors affect the choice of decision-making scheme, often there are gray factors and gray process, which is especially obvious in the fishery production system, for example, how the marine environment factors affect the change of the fishery resources, how they affect the fishery production and the central fishery, thus affecting the decision-making of the fishery production, so the Gray system theory is applied to the decision-making, this problem has been well solved. Gray decision-making is one of the important contents of gray system theory. In this chapter, the basic concept of gray decision-making, gray correlation decision-making, gray

---

X. Chen (✉)
College of Marine Sciences, Shanghai Ocean University, Lingang New City, Shanghai, China
e-mail: xjchen@shou.edu.cn

X. Chen (ed.), *Application of Gray System Theory in Fishery Science*,
https://doi.org/10.1007/978-981-99-0635-2_7

situation decision-making, and the application of gray decision-making in fishery science will be introduced.

Keywords Gray system · Decision-making · Fishery science · Gray correlation decision-making · Gray situation decision-making

7.1 Basic Concepts of Gray Decision-Making

Decision-making is one of the basic activities of human society, politics, and economic life. What is decision-making? People often interpret it as "deciding policy" or "finalizing the case" in a narrow sense. In fact, as a specific term, decision-making has very broad meaning. Any decision to act based on a predetermined goal can be called decision-making. More precisely, decision-making is a decision made on the direction, principle, and goal of future practice, as well as the methods and means to adhere to the direction, implement the principle, and achieve the goal.

A decision-making process can be roughly divided into three basic steps (Chen 2003, 2023):

The first step is to identify the problem and propose decision-making goals. Without goals, there is no decision-making. Due to the complexity of objective things, the goal is often not one but the integration and coordination of several goals. At the same time, there are contradictions between different goals. Therefore, special attention should be given to the handling, coordination, and integration of multiobjective problems. The processing methods are as follows: subordinate targets are eliminated, similar targets are merged, secondary targets are reduced to constraint conditions, and various comprehensive methods of targets and layered screening and elimination methods of targets are used. For example, in the management of fishery resources, currently, the maximum sustainable yield (MSY), the maximum economic yield (MEY), and the optimal yield (OY) are usually used as the management objectives, but these objectives are often contradictory, so each country should adapt to local conditions. Therefore, each country should formulate management objectives based on local conditions and coordinate and unify different objectives.

In the second step, through data collection, analysis and prediction research, design plans, and technical evaluation, various possible alternative plans are drawn up.

The third step is to select the most suitable scheme from various possible schemes. Based on the evaluation and analysis of various schemes, the decision is made decisively using the scientific method of selection according to certain decision-making criteria.

7.2 Gray Correlation Decision-Making

7.2.1 Basic Concepts

Now we have $S = \{s_{ij}| \ a_i \in A, b_j \in B\}$ is the situation set, $u_{i_0j_0} = \left(u_{i_0j_0}^{(1)}, u_{i_0j_0}^{(2)}, \ldots, u_{i_0j_0}^{(s)}\right)$ is the optimal effect vector. If $u_{i_0j_0}$ is the corresponding situation $u_{i_0j_0} \notin S$, then we call $u_{i_0j_0}$ the ideal optimal effect vector, and its corresponding $S_{i_0j_0}$ is called the ideal optimal situation.

Now, the effect vector corresponding to situation S_{ij} is

$$u_{ij} = \left(u_{ij}^{(1)}, u_{ij}^{(2)}, \ldots, u_{ij}^{(s)}\right); i = 1, 2, \ldots, n; j = 1, 2, \ldots, m$$

1. When the target effect value of k is larger, the value is better, $u_{i_0j_0}^{(k)} = \max_{\substack{1 \le i \le n \\ 1 \le j \le m}} \left\{u_{ij}^{(k)}\right\}$.
2. When the target effect value of k is close to a certain moderate value u_0, which is good, it is taken as good, $u_{i_0j_0}^{(k)} = u_0$.
3. When the target effect value of k is smaller, the value is better, and it is taken as $u_{i_0j_0}^{(k)} = \min_{\substack{1 \le i \le n \\ 1 \le j \le m}} \left\{u_{ij}^{(k)}\right\}$.

Then, $u_{i_0j_0} = (u_{i_0j_0}^{(1)}, u_{i_0j_0}^{(2)}, \ldots, u_{i_0j_0}^{(s)})$ is the ideal optimal effect vector.

Let $S = \{s_{ij}| \ a_i \in A, b_j \in B\}$ be the situation set, and the effect vector corresponding to situation S_{ij} is

$$u_{ij} = \left(u_{ij}^{(1)}, u_{ij}^{(2)}, \ldots, u_{ij}^{(s)}\right); i = 1, 2, \ldots, n; j = 1, 2, \ldots, m$$

$u_{i_0j_0} = \left(u_{i_0j_0}^{(1)}, u_{i_0j_0}^{(2)}, \ldots, u_{i_0j_0}^{(s)}\right)$ is the ideal optimal effect vector.

ε_{ij} $(i = 1, 2 \ldots, n; j = 1, 2 \ldots, m)$ is the gray absolute correlation degree between u_{ij} and $u_{i_0j_0}$. If $\varepsilon_{i_1j_1}$ satisfies for any $i \in \{1, 2, \ldots, n\}$ and $i \neq i_1$ and any $j \in \{1, 2, \ldots, m\}$ and $j \neq j_1$, there is always $\varepsilon_{i_1j_1} \geq \varepsilon_{ij}$, then $u_{i_1j_1}$ is the suboptimal effect vector, and $S_{i_1j_1}$ is a suboptimal situation.

7.2.2 Basic Steps of Gray Correlation Decision-Making Calculation

Gray correlation decision-making can be carried out according to the following steps (Chen 2003, 2023):

Step 1: Determine the event set $A = \{a_1, a_2, \ldots, a_n\}$ and the game set B $\{b_1, b_2, \ldots, b_m\}$, and construct the situation set $S = \{s_{ij} = (a_i\ ,\ \ b_i)|\ a_i \in A, b_i \in B\}$.

Step 2: Determine decision-making goals 1, 2, …, s.

Step 3: Find the effect value of different situations s_{ij} $(i = 1, 2\ldots, n; j = 1, 2\ldots, m)$ under the k target $u_{ij}^{(k)}$

$$u^{(k)} = \left(u_{11}^{(k)}, u_{12}^{(k)}, \ldots, u_{1m}^{(k)}; u_{21}^{(k)}, u_{22}^{(k)}, \ldots, u_{2m}^{(k)}; u_{n1}^{(k)}, u_{n2}^{(k)}, \ldots, u_{nm}^{(k)}\right); k = 1, 2, \ldots, s.$$

Step 4: Find the mean image of the situation effect sequence $u^{(k)}$ under the k target, which is still denoted as

$$u^{(k)} = \left(u_{11}^{(k)}, u_{12}^{(k)}, \ldots u_{1m}^{(k)}; u_{21}^{(k)}, u_{22}^{(k)}, \ldots u_{2m}^{(k)}; u_{n1}^{(k)}, u_{n2}^{(k)}, \ldots, u_{nm}^{(k)}\right); k = 1, 2, \ldots, s.$$

Step 5: Write the effect vector of situation s_{ij} through the fourth step.

$$u_{ij} = \left(u_{ij}^{(1)},\ u_{ij}^{(2)}, \ldots, u_{ij}^{(s)}\right); i = 1, 2, \ldots, n; j = 1, 2, \ldots, m$$

Step 6: Find the idealized optimal effect vector.

$$u_{i_0j_0} = \left(u_{i_0j_0}^{(1)}, u_{i_0j_0}^{(2)}, \ldots, u_{i_0j_0}^{(s)}\right)$$

Step 7: Calculate the gray absolute correlation degree ε_{ij} between u_{ij} and $u_{i_0j_0}$, $i = 1, 2, \ldots, n; j = 1, 2, \ldots, m$.

Step 8: From $\max\limits_{\substack{1 \le i \le n \\ 1 \le j \le m}} \{\varepsilon_{ij}\} = \varepsilon_{i_1j_1}$, we obtain suboptimal effect vector $u_{i_1j_1}$ and suboptimal situation $s_{i_1j_1}$.

7.3 Decision-Making in the Gray Situation

Situational decision-making refers to the whole process of selecting the best for a certain target under the premise of the unification of events, countermeasures, and effects. When the event and the countermeasure are quantified and the event and the decision form a paired combination of decision-making, it is called the gray situation decision-making method. If a single target is considered in decision-making, we call it single-target situation decision-making; if multiple targets are considered at the same time and the effect of the same situation under different targets is different, comprehensive analysis and overall consideration are needed to integrate these targets into one. Decision-making is called multiobjective gray situation decision-making.

Due to the incommensurability and contradiction between multiple objectives, it is impossible to simply merge multiple objectives into a single objective for decision-making. Gray system theory uses methods such as effect measurement calculation and target weighted averaging to convert multiobjective decision-making into single-objective decision-making. Circumstance restrictions. Therefore, multiobjective gray situation decision-making is a gray decision-making method with a wide range of adaptations (Chen 2003, 2023). The mathematical method for multiobjective situation decision-making is as follows:

7.3.1 Decision Element, Decision Vector, and Decision Matrix

The event is denoted as a_i, the countermeasure is b_j, and its binary combination (a_i, b_j) is called the situation and is denoted as $S_{ij} = (a_i, b_j)$. Its meaning is the jth countermeasure (b_j) to address the situation of the ith event (a_i).

$$\frac{r_{ij}}{S_{ij}} = \frac{r_{ij}}{(a_i, b_j)}$$

If there are events $a_1, a_2, \ldots, a_n$, and there are countermeasures $b_1, b_2, \ldots, b_m$, then for the same event a_i, we can use $b_1, b_2, \ldots, b_m$, etc., m countermeasures to deal with, thus forming (a_i, b_1), (a_i, b_2) ... (a_i, b_m) and m other situations. The decision-making elements corresponding to these situations can be arranged in a row to form a decision-making row:

$$\delta_i = \left[\frac{r_{i1}}{S_{i1}}, \frac{r_{i2}}{S_{i2}}, \ldots, \frac{r_{im}}{S_{im}}\right]$$

where r_{ij} is the measure of the effect of situation S_{ij}.

Similarly, for countermeasure b_i, we can use a_1, a_2 ..., a_n to match, and the corresponding decision elements can be arranged in a row to form a decision column:

$$\theta_i = \left[\frac{r_{1j}}{S_{1j}}, \frac{r_{2j}}{S_{2j}}, \ldots, \frac{r_{nj}}{S_{nj}}\right]$$

Arrange the decision rows ($i = 1, 2, \ldots, n$) and the decision columns θ_j ($j = 1, 2, \ldots, m$) to form the matrix M.

$$M = \begin{pmatrix} \frac{r_{11}}{S_{11}} & \frac{r_{12}}{S_{12}} & \cdots & \frac{r_{1m}}{S_{1m}} \\ \frac{r_{21}}{S_{21}} & \frac{r_{22}}{S_{22}} & \cdots & \frac{r_{2m}}{S_{2m}} \\ \frac{r_{n1}}{S_{n1}} & \frac{r_{n2}}{S_{n2}} & \cdots & \frac{r_{nm}}{S_{nm}} \end{pmatrix}$$

Then, M is called the situation decision matrix, which can be denoted as $M(\delta_i, \Theta_j)$.

7.3.2 *Effectiveness Measure*

The effect measurement is the measurement of the actual effect produced by the situation compared to the target. The time series is the correlation coefficient of two comparison series at the same time. The calculation formula is

$$\gamma_{ij}(t) = \frac{\Delta_{\min} + \Delta_{\max}}{\Delta_{ij}(t) + \Delta_{\max}}$$

where $\Delta_{\min}$, $\Delta_{\max}$ is the minimum difference and maximum difference (absolute value) of the subtraction of the two comparison series at each time.

$\Delta_{ij}(t)$ is the difference at any time t.

For a single point, the effect measure can be divided into (Chen 2003, 2023):

1. Upper limit effect measurement. The calculation formula is

$$\gamma_{ij} = \frac{u_{ij}}{u_{\max}}$$

$$u_{ij} \leq u_{\max}$$

where u_{ij} is the measured effect of situation S_{ij}; $u_{\max}$ is the maximum value of all the measured effects of situation S_{ij}.

$$\gamma_{ij} \leq 1$$

2. Measurement of the lower limit effect. The calculation formula is

$$\gamma_{ij} = \frac{u_{\min}}{u_{ij}}$$

$$u_{ij} \geq u_{\min}$$

where $u_{\min}$ is the minimum value of all the measured effects of situation S_{ij}.

$$\gamma_{ij} \leq 1$$

3. Central effect measurement. The calculation formula is

$$\gamma_{ij} = \frac{\min\{u_{ij}, u_o\}}{\max\{u_{ij}, u_o\}}$$

where u_0 is the reference point of the sample.

Or use the formula:

$$\gamma_{ij} = \frac{u_o}{|u_{ij} - u_0| + u_0}$$

$$\gamma_{ij} \leq 1$$

In practical applications, which effect measurement is used depends on the nature of the target. For example, for efficiency indicators such as yield and output value, the larger the better, the upper limit effect measurement is used. For example, for cost-based indicators such as investment and cost, the smaller the better, the lower limit effect measurement is used. For example, for indicators such as fertilization and irrigation, the appropriate amount is appropriate.

In addition, the steady-state effect measurement of the system can also be used for the time series of the benefits of the situation. That is, the GM (1, 1) model is established for the time series to obtain the parameters.$\widehat{a} = \begin{bmatrix} a \\ u \end{bmatrix}$ Then, $\frac{1}{a}$ is the measure of the steady-state effect of the system, denoted as:

$$\gamma_{ij} = \frac{1}{a}$$

7.3.3 Multiobjective Decision Matrix

When the situation has several targets, then the effect measure of the Kth target is recorded as $\gamma_{ij}{}^{(K)}$, and its corresponding decision element is $\frac{\gamma_{ij}{}^{(K)}}{S_{ij}}$, for which there is a corresponding decision vector $\left(\delta_i^{(K)}, \theta_{ji}^{(K)}\right)$ and decision matrix $M^{(K)}$

$$M^{(K)} = \begin{pmatrix} \frac{r^{(K)}{}_{11}}{S_{11}} & \frac{r^{(k)}{}_{12}}{S_{12}} & \cdots & \frac{r^{(K)}{}_{1m}}{S_{1m}} \\ \frac{r^{(K)}{}_{21}}{S_{21}} & \frac{r^{(K)}{}_{22}}{S_{22}} & \cdots & \frac{r^{(K)}{}_{2m}}{S_{2m}} \\ \frac{r^{(K)}{}_{n1}}{S_{n1}} & \frac{r^{(K)}{}_{n2}}{S_{n2}} & \cdots & \frac{r^{(K)}{}_{nm}}{S_{nm}} \end{pmatrix}$$

Then, the comprehensive multiobjective situation decision matrix $M^{(\Sigma)}$ is

$$M^{(\Sigma)} = \begin{pmatrix} \frac{r^{(\Sigma)}{}_{11}}{S_{11}} & \frac{r^{(\Sigma)}{}_{12}}{S_{12}} & \cdots & \frac{r^{(\Sigma)}{}_{1m}}{S_{1m}} \\ \frac{r^{(\Sigma)}{}_{21}}{S_{21}} & \frac{r^{(\Sigma)}{}_{22}}{S_{22}} & \cdots & \frac{r^{(\Sigma)}{}_{2m}}{S_{2m}} \\ \frac{r^{(\Sigma)}{}_{n1}}{S_{n1}} & \frac{r^{(\Sigma)}{}_{n2}}{S_{n2}} & \cdots & \frac{r^{(\Sigma)}{}_{nm}}{S_{nm}} \end{pmatrix}$$

The elements in the matrix are calculated as follows: $\gamma_{ij}{}^{(\Sigma)} = \frac{\sum_{K=1}^{N} \gamma_{ij}^{(K)}}{N}$.

When the K targets are biased according to the purpose and requirements of the decision-making, different weights ω_p can be given, and the elements of the matrix can be calculated as follows:

$$\gamma_{ij}{}^{(\Sigma)} = \sum_{P=1}^{N} \omega_P \gamma_{ij}^{(K)}$$

7.3.4 Decision-Making Criteria

Decision-making selects the best situation. If the best countermeasure is selected by the event, the "line decision" is performed, while the "column decision" is performed by the countermeasure matching the most suitable event. The method is

Row decision: for the decision matrix $M^{(\Sigma)}$ in the decision row δ_i, the decision element with the largest effect of the countermeasure is obtained:

$$\gamma_{ij*1}^{(\Sigma)} = \max_j \left\{ \gamma_{ij}^{(\Sigma)} \right\} = \max \left\{ \gamma_{i1}^{(\Sigma)}, \gamma_{i2}^{(\Sigma)}, \ldots, \gamma_{im}^{(\Sigma)} \right\}$$

Then, $\frac{\gamma_{ij*}^{(\Sigma)}}{S_{ij*}}$ is the decision element, $S_{ij}{}^{*}$ is the optimal decision-making situation, i.e., b_j* is the optimal countermeasure for event a_i.

Column game: For the decision matrix $M^{(\Sigma)}$ in the decision column θ_j, the decision element with the largest effect of the countermeasure is obtained:

$$\gamma_{i*j}^{(\Sigma)} = \max_i \left\{\gamma_{ij}^{(\Sigma)}\right\} = \max \left\{\lambda_{1j}^{(\Sigma)}, \gamma_{2j}^{(\Sigma)}, \ldots \gamma_{nj}^{(\Sigma)}\right\}^T$$

Then, $\frac{\gamma_{i*j}^{(\Sigma)}}{S_{i*j}}$ is called the column decision element. S_{i*j} is the optimal decision-making situation, i.e., a_i* is the optimal countermeasure for event a_i.

In the actual decision-making process, according to the above criteria, the row decision and column decision are made on the matrix, and the resulting decision is often difficult to develop in a coordinated manner in the overall situation, so the goal of the overall benefit cannot be achieved. In this case, the comprehensive matrix needs to be adjusted. Gray target decision-making can be performed after optimization or normalization.

7.3.5 Prioritized Decision Matrix

The optimization of the comprehensive decision matrix $M^{(\Sigma)}$ can be performed in two steps:

The first step is to arrange the decision elements in the matrix row by row from left to right according to the size to obtain the comprehensive decision matrix $M^{(\Sigma)}$ and the matrix $M_1^{(\Sigma)}$ after the row optimization:

$$M^{(\Sigma)} = \begin{pmatrix} \frac{0.472}{S_{11}} & \frac{0.240}{S_{12}} & \frac{0.069}{S_{13}} & \frac{0.085}{S_{14}} \\ \frac{0.416}{S_{21}} & \frac{0.157}{S_{22}} & \frac{0.067}{S_{23}} & \frac{0.268}{S_{24}} \\ \frac{0.40}{S_{31}} & \frac{0.321}{S_{32}} & \frac{0.042}{S_{33}} & \frac{0.136}{S_{34}} \\ \frac{0.382}{S_{41}} & \frac{0.449}{S_{42}} & \frac{0.077}{S_{43}} & \frac{0.068}{S_{44}} \\ \frac{0.573}{S_{51}} & \frac{0.278}{S_{52}} & \frac{0.107}{S_{43}} & \frac{0.196}{S_{54}} \\ \frac{0.331}{S_{61}} & \frac{0.193}{S62} & \frac{0.062}{S_{63}} & \frac{0.052}{S_{64}} \\ \frac{0.393}{S_{71}} & \frac{0.329}{S_{72}} & \frac{0.090}{S_{73}} & \frac{0.083}{S_{74}} \end{pmatrix}$$

$$M_1^{(\Sigma)} = \begin{pmatrix} \frac{0.472}{S_{11}} & \frac{0.240}{S_{12}} & \frac{0.085}{S_{14}} & \frac{0.069}{S_{13}} \\ \frac{0.416}{S_{21}} & \frac{0.268}{S_{24}} & \frac{0.157}{S_{22}} & \frac{0.067}{S_{23}} \\ \frac{0.40}{S_{31}} & \frac{0.321}{S_{32}} & \frac{0.136}{S_{34}} & \frac{0.042}{S_{33}} \\ \frac{0.449}{S_{42}} & \frac{0.382}{S_{41}} & \frac{0.077}{S_{43}} & \frac{0.068}{S_{44}} \\ \frac{0.573}{S_{51}} & \frac{0.278}{S_{52}} & \frac{0.196}{S_{54}} & \frac{0.107}{S_{53}} \\ \frac{0.331}{S_{61}} & \frac{0.193}{S_{62}} & \frac{0.062}{S_{63}} & \frac{0.052}{S_{64}} \\ \frac{0.393}{S_{71}} & \frac{0.329}{S_{72}} & \frac{0.090}{S_{73}} & \frac{0.083}{S_{74}} \end{pmatrix}$$

Step 2: Next, the row-optimized ordering matrix is used to arrange the decision elements from top to bottom in order of magnitude to obtain the matrix $M_2^{(\Sigma)}$, which is as follows:

$$M_2^{(\Sigma)} = \begin{pmatrix} \frac{0.573}{S_{51}} & \frac{0.382}{S_{41}} & \frac{0.196}{S_{54}} & \frac{0.107}{S_{53}} \\ \frac{0.472}{S_{11}} & \frac{0.329}{S_{72}} & \frac{0.157}{S_{22}} & \frac{0.083}{S_{74}} \\ \frac{0.449}{S_{42}} & \frac{0.321}{S_{32}} & \frac{0.136}{S_{34}} & \frac{0.069}{S_{13}} \\ \frac{0.416}{S_{21}} & \frac{0.278}{S_{52}} & \frac{0.090}{S_{73}} & \frac{0.068}{S_{44}} \\ \frac{0.40}{S_{31}} & \frac{0.268}{S_{24}} & \frac{0.085}{S_{14}} & \frac{0.067}{S_{23}} \\ \frac{0.393}{S_{71}} & \frac{0.240}{S_{12}} & \frac{0.077}{S_{43}} & \frac{0.052}{S_{64}} \\ \frac{0.331}{S_{61}} & \frac{0.193}{S_{62}} & \frac{0.062}{S_{63}} & \frac{0.042}{S_{33}} \end{pmatrix}$$

Then, we perform another check by row. If row optimization is not achieved, row permutation can be performed again. In this way, the decision-making matrix is gradually reduced from the upper left corner to form the optimal matrix M^*.

According to the optimal ordering decision matrix, the specific principles and methods for gray situation decision-making are as follows:

1. The optimal ordering matrix is divided into several steps along the main diagonal direction so that the values of the elements of the previous step are all greater than the values of the elements of the next step. The above example can be divided into the following five steps.

2. Under normal circumstances, the situation within the same rung can be selected according to the same advantages and disadvantages; that is, the situation in the upper rung is better than the situation in the lower rung and vice versa.
3. The order of preference should be performed from top to bottom, step by step. If necessary, the effect measures were compared.

$$
\begin{array}{c}
\quad\quad\quad\text{I} \quad\quad\quad\quad\quad\quad \text{II} \quad\quad\quad \text{III} \\
M^* = \begin{pmatrix}
\frac{0.573}{S_{51}} & \frac{0.382}{S_{41}} & \frac{0.196}{S_{54}} & \frac{0.107}{S_{53}} \\
\frac{0.472}{S_{11}} & \frac{0.329}{S_{72}} & \frac{0.157}{S_{22}} & \frac{0.083}{S_{74}} \\
\frac{0.449}{S_{42}} & \frac{0.321}{S_{32}} & \frac{0.136}{S_{34}} & \frac{0.069}{S_{13}} \\
\frac{0.416}{S_{21}} & \frac{0.278}{S_{52}} & \frac{0.090}{S_{73}} & \frac{0.068}{S_{44}} \\
\frac{0.40}{S_{31}} & \frac{0.268}{S_{24}} & \frac{0.085}{S_{14}} & \frac{0.067}{S_{23}} \\
\frac{0.393}{S_{71}} & \frac{0.240}{S_{12}} & \frac{0.077}{S_{43}} & \frac{0.052}{S_{64}} \\
\frac{0.331}{S_{61}} & \frac{0.193}{S_{62}} & \frac{0.062}{S_{63}} & \frac{0.042}{S_{33}}
\end{pmatrix}
\begin{array}{l} \\ \text{IV} \\ \\ \text{V} \\ \\ \text{VI} \\ \end{array}
\end{array}
$$

4. The principle of comprehensive coordination should be adhered to in decision-making. According to the specific purpose and requirements, within the same ladder, each event can choose 2–4 countermeasures, and each countermeasure can match 2–4 events to avoid the "one size fits all" in the selection of the best and prevent one event from selecting all countermeasures or one countermeasure to match all events.
5. The principle of combining qualitative and quantitative decisions should be adhered to. An echelon is similar to a "gray target," in which a satisfactory situation can be found from a nonoptimal situation, and a coordinated situation can be found from a noninferior situation.

7.3.6 Normalized Decision Matrix

For the comprehensive decision matrix $M^{(\Sigma)}$, the normalization transformation can be divided into two categories. First, the normalization process is performed row by row, and the calculation formula is

$$\gamma'_{ij} = \frac{\gamma_{ij}}{\sum_{j=1}^{m} \gamma_{ij}} \quad i = 1, 2, \ldots n$$

In this way, the row-normalized matrix is obtained $M_1^{(\Sigma)}$. This matrix can reflect the proportion of each countermeasure in the comprehensive effect measurement of each event.

The other is to perform normalization column by column, and the calculation formula is

$$\gamma'_{ij} = \frac{\gamma_{ij}}{\sum_{i=1}^{n} \gamma_{ij}} \quad j = 1, 2, \ldots, m$$

In this way, the column normalization matrix is obtained $M_2^{(\Sigma)}$. This matrix reflects the proportion of each event in the comprehensive effect measurement of each countermeasure.

For the above example matrix $M^{(\Sigma)}$ after row and column normalization, the matrices $M_1^{(\Sigma)}$ and $M_2^{(\Sigma)}$ are obtained.

$$M_1{}^{(\Sigma)} = \begin{pmatrix} \frac{0.545}{S_{11}} & \frac{0.277}{S_{12}} & \frac{0.080}{S_{13}} & \frac{0.098}{S_{14}} \\ \frac{0.458}{S_{21}} & \frac{0.173}{S_{22}} & \frac{0.074}{S_{23}} & \frac{0.295}{S_{24}} \\ \frac{0.445}{S_{31}} & \frac{0.357}{S_{32}} & \frac{0.047}{S_{33}} & \frac{0.151}{S_{34}} \\ \frac{0.391}{S_{41}} & \frac{0.460}{S_{42}} & \frac{0.079}{S_{43}} & \frac{0.070}{S_{44}} \\ \frac{0.497}{S_{51}} & \frac{0.240}{S_{52}} & \frac{0.093}{S_{53}} & \frac{0.170}{S_{54}} \\ \frac{0.519}{S_{61}} & \frac{0.303}{S_{62}} & \frac{0.097}{S_{63}} & \frac{0.081}{S_{64}} \\ \frac{0.439}{S_{71}} & \frac{0.368}{S_{72}} & \frac{0.100}{S_{73}} & \frac{0.093}{S_{74}} \end{pmatrix}$$

$$M_2^{(\Sigma)} = \begin{pmatrix} \frac{0.159}{S_{11}} & \frac{0.122}{S_{12}} & \frac{0.134}{S_{13}} & \frac{0.096}{S_{14}} \\ \frac{0.140}{S_{21}} & \frac{0.080}{S_{22}} & \frac{0.130}{S_{23}} & \frac{0.302}{S_{24}} \\ \frac{0.135}{S_{31}} & \frac{0.163}{S_{32}} & \frac{0.082}{S_{33}} & \frac{0.153}{S_{34}} \\ \frac{0.125}{S_{41}} & \frac{0.228}{S_{42}} & \frac{0.150}{S_{43}} & \frac{0.077}{S_{44}} \\ \frac{0.193}{S_{51}} & \frac{0.141}{S_{52}} & \frac{0.208}{S_{53}} & \frac{0.221}{S_{54}} \\ \frac{0.122}{S_{61}} & \frac{0.098}{S_{62}} & \frac{0.121}{S_{63}} & \frac{0.059}{S_{64}} \\ \frac{0.132}{S_{71}} & \frac{0.167}{S_{72}} & \frac{0.175}{S_{73}} & \frac{0.093}{S_{74}} \end{pmatrix}$$

Using the two normalized matrices obtained, the gray situation decision can be made. The specific method is as follows:

1. Normalize the matrix with columns $M_2^{(\Sigma)}$ to carry out decision-making and select the best decision for each event, and the optimal situation will be found. In the above example, the optimal positions of each row are S_{11}, S_{24}, S_{32}, S_{42}, S_{54}, S_{61}, and S_{73}.
2. Use the row-normalized matrix $M_1^{(\Sigma)}$ to select the best matching event of each countermeasure and form the optimal situation. In the above example, the optimal situation of each column is S_{11}, S_{42}, S_{24}, and S_{73}.
3. Following the above two steps and selecting the suboptimal (or satisfactory) situation. For example, S_{13}, S_{21}, S_{34}, S_{43}, S_{53}, S_{63}, S_{24}, and S_{72} are in the column normalization matrix, and S_{61}, S_{72}, S_{63}, and S_{54} are in the row normalization matrix.
4. On the basis of the above results, global coordination is performed. That is, the row normalization is coordinated, the row and column of the column normalization matrix are coordinated, and the situations of "globally superior and local nonoptimal" and "globally nonoptimal and locally superior" are checked and adjusted. For example, in situation S_{51}, in the column-normalized matrix, the global is optimal (the proportion of the column is the largest), but the horizontal comparison is nonoptimal in the row-normalized matrix, and it is also locally optimal in the row-normalized matrix (according to the row-based normalization matrix). However, the vertical comparison is not optimal according to the column, so the decision is lost, and S_{51} should be re-elected during coordination. S_{53} should be screened out for similar analysis.

7.4 Analysis of the Application of the Gray Decision-Making System in Fishery Science

7.4.1 Application in the Fish Farming Industry

In the aquaculture industry, there are many factors that restrict the production of fish farming, including feed factors and environmental factors, as well as the interspecific relationships among the cultured species. To scientifically and rationally develop the aquaculture industry, protect the aquatic environment, and improve the efficiency of aquaculture, it is necessary to make scientific decisions on the factors and systems that affect the fish aquaculture industry. Some scholars have used the gray decision system to conduct research in this area and have achieved some results.

Xie et al. (1998) published an article entitled "The comparative study on factors analysis and yield model of high-yield fish-pond for the Pearl River Delta and Yangtze Delta." In this study, we collected the relevant data of the "Comprehensive High-yield Technology Experiment of Ten Thousand Mu of Continuous Fish Ponds in the Pearl River Delta Region" in the Shunde area in 1983, including the net yield X_0, the stocking amount of bighead carp X_1, the stocking amount of silver carp X_2, the stocking amount of grass carp X_3, the stocking amount of mud carp X_4, the stocking amount of trash fish X_5, the protein content of the concentrate X_6, and the protein content of forage X_7.

In this study, the gray correlation decision-making method was used to analyze the correlation degree of the factors affecting the yield, and the data were standardized using the mean method with a resolution coefficient of 0.5. The results were as follows:

The gray correlation between the stocking amount of bighead carp and the net yield was $r_1 = 0.657$.

The gray correlation between the stocking amount and the net yield of silver carp was $r_2 = 0.474$.

The gray correlation between the stocking amount of grass carp and the net yield was $r_3 = 0.599$.

The gray correlation degree between the stocking amount of carp and the net yield was $r_4 = 0.709$.

The gray correlation degree of trash fish stocking amount and net yield was $r_5 = 0.489$.

The gray correlation degree of concentrate protein amount and net yield was $r_6 = 0.717$.

The gray correlation degree of forage protein amount and net yield was $r_7 = 0.762$.

According to the gray correlation degree, the gray correlation sequence is as follows: $r_7 > r_6 > r_4 > r_1 > r_3 > r_5 > r_2$.

The above results indicate that the factors that have a greater impact on net yield are the forage protein content, concentrate protein content, and stocking amount of

Table 7.1 Dominance analysis matrix of high-yield ponds in the Shunde area (Xie et al. 1998)

Content	Net yield of grass carp (X_1)	Net yields of silver carp (X_2)	Net yields of mup carp (X_3)	Net yields of bighead carp (X_4)
Fishing stocks of grass carp (Y_1)	1.0000 (r_{11})	0.8242 (r_{12})	0.4549 (r_{13})	0.2458 (r_{14})
Fishing stocks of silver carp (Y_2)	0.5431 (r_{21})	1.0000 (r_{22})	0.4625 (r_{23})	0.2209 (r_{24})
Fishing stocks of mud carp (Y_3)	0.4759 (r_{31})	0.6937 (r_{32})	1.0000 (r_{33})	0.2102 (r_{34})
Fishing stocks of bighead carp (Y_4)	0.8075 (r_{41})	0.7086 (r_{42})	0.5671 (r_{43})	1.0000 (r_{44})

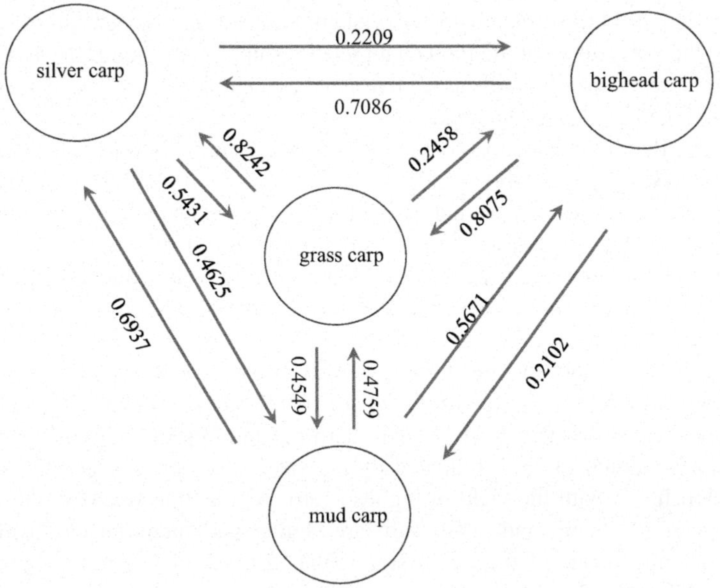

Fig. 7.1 Gray correlation degree of the polyculture fish relationship (Xie et al. 1998)

mud carp, followed by the stocking amount of bighead carp and grass carp, and the least important factor is the stocking amount of trash fish and silver carp.

The study also used the collected data to analyze the dominant factors of the high-yield ponds in the Shunde area. Using the stocking amounts of the four fish species as the reference series, Y_1, Y_2, Y_3, and Y_4 represented the fishing stocks of grass carp, silver carp, mud carp, and bighead carp per 1/5/ha, respectively. The net yields of grass carp, silver carp, mud carp, and bighead carp per ha were denoted as X_1, X_2, X_3, and X_4, respectively, with a resolution coefficient of 0.1. The analysis results are shown in Table 7.1 and Fig. 7.1.

Based on the above analysis, the study concluded the following:

1. Forage protein and concentrate protein have the greatest impact on net yield.
2. The effect of the same parent factor (Y) on different subfactors (X).

   The effect of grass carp stocking (Y_1) on the yield of various fish species: $r_{(11)} =$ (gray correlation between Y_1 and X_1) = (grass carp stocking amount to net yield of grass carp) = 1.000, $r_{(12)} =$ (grass carp stocking to net yield of sliver carp) = 0.8242, $r_{(13)} =$ (grass carp stocking to net yield of mud carp) = 0.4549, $r_{(14)} =$ (grass carp stocking to net yield of bighead carp) = 0.2458. This indicates that the stocking amount of grass carp has a greater impact on the yield of silver carp, which is consistent with the relationship that we usually think of as "three silver carp in one grass belt." At the same time, we also know that $r_{(12)}$ is the largest value in the matrix, and the main basis of multispecies polyculture in China is to use grass carp culture as the main management object. The stocking of grass carp has little effect on the net yield of bighead carp, which can still be seen from the perspective of the food chain relationship. That is, mud carp is a benthic fish that feeds mainly on benthic phytoplankton, similar to the diet of silver carp, but bighead carp is a zooplankton.

   The effect of silver carp stocking (Y_2) on the yield of various fish species was as follows: $r_{(22)} = 1.0000 > r_{(21)} = 0.5431 > r_{(23)} = 0.4625 > r_{(24)} = 0.2209$. This indicates that the stocking amount of silver carp has the greatest impact on the net yield of grass carp. Silver carp mainly play a role in regulating water quality, while grass carp require freshwater quality to be conducive to growth. The stocking amount of silver carp was the second most important, indicating that the feed of silver carp was close to that of common carp.

   The effect of the stocking amount of mud carp (Y_3) on the yield of various fish species was as follows: $r_{(33)} = 1.000 > r_{(32)} = 06937 > r_{(31)} = 0.4759 > r_{(34)} = 0.2102$. This indicates that the yield of silver carp has the largest relationship with the stocking amount of carp, followed by the yield of grass carp, and the smallest relationship is with the yield of bighead carp. As mentioned above, the feeding habits of mud carp and silver carp are similar and therefore closely related. The feeding habits of mud carp are similar to those of grass carp fingerlings. The relationship between mud carp and bighead carp is basically irrelevant, so the correlation coefficient is the smallest in the entire matrix.

   The effect of bighead carp stocking (Y_4) on the yield of various fish species was as follows: $r_{(44)} = 1.000 > r_{(41)} = 0.8075 > r_{(42)} = 0.7086 > r_{(43)} = 0.5671$. This indicates that the stocking amount of bighead carp has the largest relationship with the net yield of grass carp, and it is second in the entire matrix. Due to the presence of mud carp in the pond, the water quality is relatively fat. The relationship between silver carp and bighead carp was second, and the relationship between bighead carp and mud carp was the smallest.
3. Effect of different parent factors (Y) on the same subfactor (X)

   $r_{(11)} =$ (degree of gray correlation between Y_1 and X_1) = (grass carp stocking to net grass carp yield) = 1.000, $r_{(21)} =$ (silver carp stocking to net grass carp yield) = 0.5431, $r_{(31)} =$ (the stocking amount of bighead carp to the net yield of grass carp) = 0.4759, $r_{(41)} =$ (the stocking amount of bighead carp to the net yield of grass carp) = 0.8075. Therefore, $r_{(11)} > r_{(41)} > r_{(21)} > r_{(31)}$. The results showed that the first major factor affecting the net yield of grass carp was the stocking of grass carp, followed by the stocking of bighead carp, and the smallest was the stocking of mud carp.

The effect of different stocking conditions on silver carp yield (X_2): $r_{(21)} = 1.000 > r_{(11)} = 0.8242 > r_{(41)} = 0.7086 > r_{(31)} = 0.6937$. This indicates that the stocking of silver carp has the greatest impact on the yield of silver carp, followed by the stocking of grass carp, and the smallest impact is the stocking of mud carp.

The effect of different stocking conditions on the yield (X_3) of mud carp was as follows: $r_{(31)} = 1.000 > r_{(41)} = 0.5671 > r_{(21)} = 0.4625 > r_{(11)} = 0.4549$. This indicates that the impact on the yield of mud carp was followed by the stocking of mud carp and the stocking of bighead carp, and the smallest impact was the stocking of grass carp.

The effect of different stocking conditions on the yield (X_4) of bighead carp was as follows: $r_{(41)} = 1.000 > r_{(11)} = 0.258 > r_{(21)} = 0.2209 > r_{(31)} = 0.2102$. Studies have shown that stocking bighead carp has the greatest impact on the yield of bighead carp, followed by stocking grass carp, and stocking mud carp has the smallest impact.

4. Advantage analysis in the matrix

It can be seen from Table 7.1 that, from the perspective of the rows of the matrix, each data point in the fourth row is greater than the corresponding data point in the other rows, that is, the stocking amount of the bighead is the dominant factor of the matrix.

The aquaculture in the Shunde area, which is located in the Pearl River Delta, is different from that in other areas. Based on the characteristics of rich water and fast growth of bighead carp in the ponds, the mixed culture of bighead carp was the main type of pond, while silver carp was less common, which did not inhibit bighead carp in food. The growth potential of bighead carp could be brought into full play, and the total yield of the pond could be increased. Therefore, the stocking of bighead carp is the dominant maternal factor. The least influence on yield was the stocking amount of mud carp, which was a nondominant factor.

From the columns in Table 7.1, each data point in the second column is greater than the corresponding data points in the other columns; that is, the net yield of silver carp is the dominant subfactor of the matrix. From the above analysis, we can see that the relationship between the yield of silver carp and the stocking of several fish species is greater than that of other fish species. First, the stocking of grass carp has the largest influence factor. Due to the importance of bighead carp culture in the Pearl River Delta, the yield of silver carp was also affected because of its close relationship with mud carp. Therefore, the net yield of silver carp was the most affected factor, and the yield of silver carp was the dominant factor. The net yield of bighead carp was the least influential factor, and the net yield of bighead carp was the nondominant factor.

7.4.2 *Application of Environmental Assessment of Fishery Waters*

The environment is a complex system with multiple factors and multiple levels. Because there is some unclear gray information in the quality assessment of the environmental system, it is difficult to establish a definitive mathematical model based on the monitoring data of the assessment factors. Therefore, it is more objective and reasonable to use gray system theory to evaluate the eutrophication of fishery waters.

Zheng and Li (1999) published "a modified gray situation decision-making method for the assessment of lake water eutrophication." An example shows that this model is applicable to lake water quality assessment, the method is feasible, and the results are reasonable.

7.4.2.1 Determine the Event Set, Countermeasure Set, and Target Set

The 9 major lakes (Table 7.2) constitute the event set $A = \{a_1, a_2, \ldots, a_9\} = \{$Qinghai Lake, Taihu Lake, . . ., Erhai Lake$\}$.

The eutrophication status of the lake water body is divided into five levels (Table 7.3). The target set $P = \{$total phosphorus..., biomass$\}$ was formed by the five pollution parameters participating in the evaluation.

Table 7.2 Measured data of evaluation parameters of nine major lakes in China (Zheng and Li 1999)

Lakes	Qinghai Lake	Taihu	Hulun Lake	Hongze Lake	Chaohu
Total phosphorus/μg·L^{-1}	20	20	80	100	30
Chemical oxygen consumption/mg·L^{-1}	1.4	2.83	8.29	5.5	6.26
Transparency/m	4.5	0.5	0.5	0.3	0.25
Total nitrogen/mg·L^{-1}	0.22	0.9	0.13	0.46	1.67
Biomass/10 thousand ind.·L^{-1}	14.6	100	11.6	11.5	25.3
Lakes	Dianchi	Wuhan East Lake	Hangzhou West Lake	Erhu	
Total phosphorus/μg·L^{-1}	20	105	130	34	
Chemical oxygen consumption/mg·L^{-1}	10.13	10.7	10.3	2.11	
Transparency/m	0.5	0.4	0.35	3.3	
Total nitrogen/mg·L^{-1}	0.23	2.0	0.76	0.49	
Biomass/10 thousand ind.·L^{-1}	189.2	1913.7	6920	22.30	

Table 7.3 Lake water quality classification standards (Zheng and Li 1999)

Level	Total phosphorus μg·L^{-1}	Chemical oxygen consumption mg·L^{-1}	Transparency m	Total nitrogen mg·L^{-1}	Biomass 10 thousand ind.·L^{-1}
Extremely poor nutrition	<1	<0.09	>37.0	<0.02	<4
Poor nutrition	4	0.36	12.0	0.06	15
Medium nutrition	23	1.80	2.4	0.31	50
Eutrophication	110	7.1	0.55	1.20	100
Very nutritious	>660	>27.10	<0.17	>4.60	>1000

7.4.2.2 Calculate the Measure of Target Effectiveness

The whitening function is used to calculate the target effect measurement. For example, the whitening function of target 1 (total phosphorus) on the water quality of the first-class lake is

$$\gamma_{i1}^{(1)} = \begin{cases} 1 & x_{i1} < 1 \\ (4 - x_{i1})/3 & 1 \leq x_{i1} \leq 4 \\ 0 & x_{i1} > 4 \end{cases}$$

For grade 2 water quality:

$$\gamma_{i2}^{(1)} = \begin{cases} (x_{i1} - 1)/3 & 1 \leq x_{i1} \leq 4 \\ (23 - x_{i1})/19 & 4 < x_{i1} \leq 23 \\ 0 & x_{i1} < 1 \quad x_{i1} > 23 \end{cases}$$

For grade 5 water quality:

$$\gamma_{i5}^{(1)} = \begin{cases} 0 & x_{i1} < 110 \\ (x_{i1} - 110)/660 & 100 \leq x_{i1} \leq 660 \\ 1 & x_{i1} > 660 \end{cases}$$

Similarly, the whitening function of each target on the eutrophication level of the five lakes can be established.

By substituting the whitening value of each target (Table 7.2) into the corresponding formula, the effect measurement of each target can be obtained, and the effect measurement matrix is formed. For example, the effect measurement matrix of target 1 (total phosphorus) is

$$R^{(1)} = \begin{vmatrix} 0 & 0 & 0 & 0 & 0 & 0 & 0 & 0 & 0 \\ 0.158 & 0.158 & 0 & 0 & 0 & 0.158 & 0 & 0 & 0 \\ 0.842 & 0.842 & 0.345 & 0.115 & 0.920 & 0.842 & 0.057 & 0 & 0.874 \\ 0 & 0 & 0.655 & 0.885 & 0.080 & 0 & 0.943 & 0.964 & 0.126 \\ 0 & 0 & 0 & 0 & 0 & 0 & 0 & 0.036 & 0 \end{vmatrix}$$

7.4.2.3 Determination of Target Weights

According to the whitening value of each target, the target weights are calculated according to the formula $\omega_{ik} = \overline{\omega}_{ik} / \sum_{k=1}^{p} \overline{\omega}_{ik}$, where the standard value of the third level (medium nutrition) is the reference value S_{0k}, and the calculation results of each target weight are listed in Table 7.4.

7.4.2.4 Calculate the Comprehensive Effect Measure

According to the effectiveness of each target $\gamma_{ij}^{(k)}$ and weight ω_{ij}, the comprehensive effect measure $\gamma_{ij}^{(\Sigma)}$ is calculated according to $\gamma_{ij}^{(\Sigma)} = \sum_{k=1}^{p} \omega_{ik} \gamma_{ij}^{(k)}$ and thus constitutes the comprehensive effect measurement matrix as follows:

Table 7.4 Calculation results of the weight of each target (Zheng and Li 1999)

Lakes	Qinghai Lake	Taihu	Hulun Lake	Hongze Lake	Chaohu
Total phosphorus	0.192	0.115	0.389	0.470	0.121
Chemical oxygen consumption	0.172	0.208	0.515	0.330	0.322
Transparency	0.414	0.028	0.023	0.014	0.010
Total nitrogen	0.157	0.384	0.047	0.161	0.500
Biomass	0.056	0.265	0.026	0.025	0.047
Lakes	Dianchi	Wuhan East Lake	Hangzhou West Lake	Erhu	
Total phosphorus	0.077	0.082	0.036	0.032	
Chemical oxygen consumption	0.501	0.107	0.036	0.025	
Transparency	0.019	0.004	0.001	0.030	
Total nitrogen	0.066	0.116	0.056	0.034	
Biomass	0.337	0.691	0.871	0.878	

$$R^{(\sum)} = \begin{vmatrix} 0.002 & 0 & 0.008 & 0.008 & 0 & 0 & 0 & 0 & 0 \\ 0.288 & 0.018 & 0.053 & 0.017 & 0.032 & 0.012 & 0 & 0 & 0.697 \\ 0.710 & 0.394 & 0.147 & 0.288 & 0.176 & 0.186 & 0.095 & 0 & 0.289 \\ 0 & 0.584 & 0.759 & 0.678 & 0.715 & 0.790 & 0.266 & 0.196 & 0.014 \\ 0 & 0.004 & 0.033 & 0.009 & 0.077 & 0.012 & 0.639 & 0.804 & 0 \end{vmatrix}$$

7.4.2.5 Determining the Optimal Situation

The formula $H_i = \sum_{j=1}^{m} j\gamma_{ij}^{(\Sigma)}$ was used to calculate the eigenvalues H_i of the grade variables of each sample and determine the degree of eutrophication of each lake.

It is concluded that $H_1 = 2.70$, $H_2 = 3.57$, $H_3 = 3.76$, $H_4 = 3.66$, $H_5 = 4.24$, $H_6 = 3.54$, $H_7 = 4.23$, $H_8 = 4.32$, and $H_9 = 2.72$.

Therefore, it was determined that Qinghai Lake was mesotrophic, Taihu Lake and Dianchi Lake were between grades 3 and 4, and they were mesotrophic.

References

Chen XJ (2003) Application of gray system theory in fishery science. China Agricultural Press. (In Chinese).

Chen XJ (2023) Application of gray system theory in fishery science. China Agricultural Press. (In Chinese)

Xie J, Xiao XZ, Huang ZH et al (1998) The comparative study on factors analysis and yield model of high-yield fish-pond for the Pearl river delta and Yangtze delta. J Shanghai Fish Univ 7(2):102–106. (In Chinese)

Zheng CD, Li ZB (1999) Improved grey situation decision making method for lake eutrophication evaluation. J Lake Sci 11(1):75–80. (In Chinese)

Chapter 8
Gray Linear Programming

Xinjun Chen

Abstract Linear programming is an important branch of operational research, which has been studied earlier, developed faster, applied widely, and used more mature methods. It is a mathematical theory and method to study the extreme value problem of linear objective function under linear constraint, which is widely used in economic analysis, management, engineering technology, industrial and agricultural production, etc., it provides a scientific basis for making optimal decision with limited human, material, and financial resources. Generally, under the condition of linear constraint, the problem of finding the maximum or minimum of linear objective function, the decision variable, constraint condition and objective function are the three elements of linear programming. However, there are many uncertainties in fisheries production and aquaculture, such as the impact of the global climate and marine environment on the early life histories of fish, the impact of fishing practices on target species and their ecosystems, and so on, these cannot be expressed by an exact value, so there are many uncertain and fuzzy factors in the social economic system and natural ecological environment system of fishery production, the phenomenon is always gray, and the wrong results may appear when the traditional linear programming is used to analyze and deal with the problem. The emergence of Gray system provides a way to solve this problem. Gray linear programming is a kind of dynamic linear programming, which can be used to solve the problems of conventional linear programming and achieved good results. In this chapter, the gray linear programming model and its calculation method are introduced, and the application of gray linear programming in fishery science is analyzed by taking the study on the adjustment of marine fishery structure in Shandong province as an example, the development trend of fisheries in Shandong province is forecasted scientifically, the power and composition of fishing vessels of different grades are calculated, and the suggestions for the adjustment of marine fisheries structure in Shandong province are put forward.

X. Chen (✉)
College of Marine Sciences, Shanghai Ocean University, Lingang New City, Shanghai, China
e-mail: xjchen@shou.edu.cn

X. Chen (ed.), *Application of Gray System Theory in Fishery Science*,
https://doi.org/10.1007/978-981-99-0635-2_8

Keywords Gray linear programming · Fishery science · Marine fishing · Mariculture

Linear programming is a decision-making method that is widely used in the study of multivariable systems, and it is particularly popular in socioeconomic disciplines and fisheries sciences. However, because there are many uncertain and fuzzy factors in the socioeconomic system, fisheries system, and the other natural ecological environment systems, their phenomenon is often gray. Therefore, errors may occur when using linear programming to analyze and deal with problems. Gray linear programming is carried out under the condition that the technical coefficient is a variable gray number and the constraint value is developed. It is a kind of dynamic linear programming that compensates for the shortcomings of conventional linear programming and has also been preliminarily applied in fishery science. This chapter mainly introduces the gray linear programming model and the application of gray linear programming in fishery science.

8.1 Gray Linear Programming Model

8.1.1 Standard Form of the Linear Programming Model

Linear programming is an important branch of operations research. It is a mathematical model that is widely used and easy to implement in the study of multivariable systems. It is also the most commonly used method for deterministic decision-making. The main problem that it solves is how to maximize the role of limited resources (including human resources) to achieve the maximum economic and social benefits and to find effective ways for the rational use of human, material, and financial resources. There are two main types of problems in the study of linear programming: first, after a goal or task is determined, how to make overall arrangements and use the least manpower, material resources and financial resources to complete this goal; second, under certain conditions including human, material, and financial resources, how to reasonably arrange and use human, material, and financial resources to achieve the most tasks and maximize the benefits. This is actually two aspects of the same problem, that is, solving the overall optimal problem of the system. Therefore, linear programming is often used as the main mathematical method to adjust the industrial structure of various industries.

Linear programming is used to solve linear relationship problems. The so-called linear relationship is a proportional relationship. For example, the relationship between production and resource input and between cost and profit is generally linear or close to linear. The following conditions are usually required to form a linear programming problem (Chen 2003, 2023):

1. Determine the decision variables of the problem. This refers to the factors that the decision maker can control, and their values determine the solution of the model.
2. There must be clear goals. It is required that the objective of the question can be expressed by a numerical value, that is, the relevant question is converted into a formula, and the criteria used by the decision maker to evaluate different answers to the question, namely the objective function, are determined.
3. The goal to be achieved is realized under certain constraints, and there are multiple feasible schemes to achieve the goal.
4. To clarify the limited number of limited resources, the input–output relationship and the output-benefit relationship of each production sector are used to determine the reasonable coefficients of the decision-making variables.
5. Both the constraint condition and the objective function must have a linear relationship. The constraint conditions reflect the limitations of the system environment, and the objective function reflects the goals of the decision makers.

Therefore, the general linear programming model includes five parts:

1. decision variable X_j ($j = 1, 2, \ldots, n$);
2. constraint or resource constraint b_i ($i = 1, 2, \ldots, n$);
3. technical coefficient a_{ij};
4. benefit coefficient c_j;
5. objective function Z.

The mathematical model of linear programming is
Objective function max or min $Z = c_1 x_1 + c_2 x_2 + \ldots\ldots + c_n x_n$
Satisfied with the constraints:

$$a_{11}x_1 + a_{12}x_2 + \cdots + a_{1n}x_n = b_1$$

$$a_{21}x_1 + a_{22}x_2 + \cdots + a_{2n}x_n = b_2$$

$$\cdots$$

$$a_{m1}x_1 + a_{m2}x_2 + \cdots + a_{mn}x_n = b_m$$

$$x_1, x_2, \ldots, n \geq 0$$

Its abbreviation is

Objective function max or min $Z = \sum_{j=1}^{n} c_j x_j$

Satisfy the constraints $\sum_{j=1}^{n} a_{ij}x_j = b_i$ ($I = 1, 2, \ldots, m$)

$$x_j \geq 0 \ (j = 1, 2, \ldots, n)$$

where x_j is a set of unknown decision variables representing the output of various products; a_{ij} is the technical coefficient, which represents the input quantity of i types of production factors required to produce j types of products; c_j is the efficiency coefficient, which represents the income of the production unit j types of products; b_i is the restricted quantities of production factors.

The linear programming problem with the above structure is called the standard form. The specific linear programming model may have many limitations and constraints, but any linear programming problem can be transformed into the above standard form.

8.1.2 Gray Linear Programming

Although linear programming has been widely used in fields of social and economic development, fisheries science, etc., general linear programming has the following problems (Chen 2003, 2023):

1. Linear programming is static and cannot reflect the change in constraint conditions over time. Therefore, the obtained results often fail due to changes in conditions.
2. If there are gray parameters (or gray numbers) in the planning model, such as the technical coefficients and constraint values in the constraint equations, it is difficult to address general linear programming.
3. Due to the problem of model technology or computational skills, there is often no solution or unsolvable problem in the actual calculation process.

Due to the above problems, the application of general linear programming is limited to a certain extent. However, these problems can be solved using the idea and modeling method of the gray system. Linear programming combined with gray system theory is called gray linear programming.

The form of gray linear programming is as follows:

Objective function: $f(X) = CX \rightarrow \max(\text{or } \min)$

Constraints: $\otimes(A)X \leq b \; X \geq 0$

In other words, satisfying $\otimes(A)X \leq b$ under the condition of $X \geq 0$, a set of X is sought to make $f(X)$ reach the maximum value (or minimum value).

In the above relation, X is a vector:

$$X = [x_1, x_2, \cdots, x_n]^T$$

C is the coefficient vector of the objective function

$$C = [c_1, c_2, \cdots, c_n]$$

where C_i can be a gray number, $\otimes(A)$ is the coefficient matrix of the constraint condition, and A is the whitening matrix of $\otimes(A)$ and has

$$\otimes(A) = \begin{pmatrix} \otimes_{11} & \otimes_{12} & \cdots & \otimes_{1n} \\ \vdots & \vdots & \cdots & \vdots \\ \otimes_{m1} & \otimes_{m2} & \cdots & \otimes_{mn} \end{pmatrix}$$

$$A = \begin{pmatrix} a_{11} & a_{12} & \cdots & a_{1m} \\ \vdots & \vdots & \cdots & \vdots \\ a_{m1} & a_{m2} & \cdots & a_{mn} \end{pmatrix}$$

b is the constraint quantity $b = [b_1, b_2, \cdots, b_m]^T$.

If there is a set of whitening sequences for the constraint index b_i, there is a set of whitening sequences.

$$b_i^{(o)} = \left\{ b_i^{(o)}(1), b_i^{(o)}(2), \cdots, b_i^{(o)}(N) \right\}$$

Then, after $b_i^{(o)}$ is accumulated to obtain $b_i^{(1)}$, $b_i^{(1)}$ is used to set up the prediction model GM (1, 1), and then the prediction value is obtained from the prediction model.

$$\left\{ b_i^{(0)}(K), K \rangle n \right\}$$

When making the planning calculation, the following constraint conditions are applied.

$$\otimes(A)\ X\ = \begin{pmatrix} b_1^{(0)}(K) \\ b_2^{(0)}(K) \\ \vdots \\ b_m^{(0)}(K) \end{pmatrix}$$

Then, the gray linear programming value at time K can be obtained. When $K > n$ is set to different values, various linear programming solutions for future development can be obtained, that is, linear programming solutions for different periods.

Gray linear programming has the following characteristics (Chen 2003, 2023):

1. It makes up for the shortcomings of general linear programming. Conventional linear programming is a deterministic and static model that requires that the benefit coefficient in the target coefficient, the technical coefficient in the constraint condition, the amount of resources, and other restrictions be fixed. In fact, the socioeconomic relationship is uncertain and changeable, and there are many accidental and risky factors. In practice, there is no solution. Gray linear programming is carried out under the condition that the technical coefficients are variable gray numbers and the constraint values are developed. It is a dynamic

linear programming that compensates for the shortcomings of conventional linear programming.

2. It can not only guide the optimal structure under given conditions but also guide the development and change of the optimal structure. The constraint values in the constraint conditions may be variable, and some can be described by time series. The GM (1,1) model is used for prediction. Such linear programming not only reflects a specific situation but can also reflect the development and change of constraints. Such a linear programming solution is not a value but a group of values and is a group of time series values. Such a solution can not only guide the optimal structure under the present condition but also provide information on the development of the optimal structure relationship.
3. Given a set of information, a set of optimization schemes can be obtained. The constraint condition coefficient in gray linear programming is the number of gray intervals, which can be planned according to the lower bound, the upper bound, or any whitening value in the interval. In the interval, as long as a set of whitening values (information) can be obtained, a set of optimization schemes can be obtained so that the planning is flexible and has much room for adjustment, adapting to the development and changes of the situation and avoiding the fact that the conventional linear programming makes a lot of specificity. No feasible solution can be obtained for the problem.

8.2 Application of Gray Linear Programming in Fishery Science

At present, the application of gray linear programming in fishery science is mainly in the aspects of marine fishery structure characteristics (including aquaculture, fishing vessel structure, etc.), industrial structure adjustment, and development planning. Gao et al. (1999) published the "Study on the structural adjustment of marine fisheries in Shandong province," which studied the application of the gray linear programming optimization method to explore the adjustment of the fishing vessel structure to achieve the optimization of the output value.

8.2.1 Analysis of the Structure of Different Fishing Vessels

1. Target determination: This study selects the indicator reflecting the economic benefits, maximum profit, as the objective function and calculates the suitable marine fishing effort of each level of fishing vessel with the maximum economic benefit within the predicted range.
2. Variable setting: The fishing effort of each level of fishing vessel is selected as the decision variable.

3. Constraint conditions: Among the many available constraint variables, total control and sublevel ship power control are used as dual constraint parameters.
4. Selection of coefficients: The benefit coefficient used in this study refers to the coefficient of the decision variable in the objective function, which refers to the profit per 1000 PS of fishing vessels.
5. Establishment of the linear programming model. In 2000, the linear programming model of marine fishing vessels in Shandong Province was as follows (Gao et al. 1999):

$$Y_{\max} = 2497X_1 + 3256X_2 + 2818X_3 + 1540X_4 + 2062X_5 + 109X_6 + 3X_7 + 398\,X_8$$

$$X_1 + X_2 + X_3 + X_4 + X_5 + X_6 + X_7 + X_8 \leq 100$$

$$X_5 + X_6 + X_7 + X_8 \leq 400$$

$$X_1 \leq 10, X_2 = 20 - 25, X_3 = 5 - 10, X_4 \leq 20$$

$$X_1, X_2, X_3, X_4, X_5, X_6, X_7, X_8 \geq 0$$

   where X_1, X_2, X_3, X_4, X_5, X_6, X_7, and X_8 are below 19 PS, 20 PS, 21–59 PS, 60–119 PS, 120–199 PS, 200–399 PS, 400–599 PS and above 600 PS fishing boat horsepower, respectively.
6. Calculation results: Since the planned value is an approximate number, the calculation results are rounded to the nearest whole number (all units are 10^4 PS).

$$X_1 = 10$$

$$X_2 = 25$$

$$X_3 = 5$$

$$X_4 = 20$$

$$X_5 = 10$$

$$X_6 = 10$$

$$X_7 = 5$$

$$X_8 = 15$$

Under the structure of fishing vessels at all levels under the condition of controlling the total fishing effort at 100×10^4 PS, the calculation results of the above optimization model show that the proportion of fishing vessels below 19 PS should be reduced, and the proportion of fishing vessels at 21–59 PS and 120–199 PS should be also reduced, the fishing vessels at 20 PS should be kept stable, and the fishing vessels at 60–119 PS, 200–399 PS, and above 600 PS should be developed (Gao et al. 1999).

8.2.2 The Planning of the Number of Fishing Vessels and Their Efforts in Marine Fishing Operations in Shandong Province

Based on the analysis of the types of fishing vessels and the allocation of efforts in Shandong Province, the following adjustments are proposed (Gao et al. 1999):

1. Under the condition that the total marine fishing effort is controlled to be less than 100×10^4 PS, considering the carrying capacity of fishery resources and the existing fishery productivity, the fishing effort structure with a 4:3:3 ratio is proposed, i.e., trawling boat (including purse seine) occupies 40% of the total, gill-net boat (including jigging boat) accounts for 30% of the total, and stake net boat (including other fishing boats) occupies 30% of the total.
2. Effort allocation for different types of operations (Table 8.1). The total number of fishing vessels is controlled at approximately 26,680, which is a significant decrease from the current 35,417. At the same time, the fishing effort and the structure of the types of fishing vessels have been significantly improved.

8.2.3 Linear Programming Model for the Marine Aquaculture Industry

The objective function is selected to reflect the economic benefit index, which is the maximum net income, to obtain the suitable aquaculture area for each industry. The aquaculture area was selected as the decision variable. The aquaculture areas of fish farming, shrimp and crab farming, algae farming, shallow sea shellfish farming, and tidal flat farming were used as the decision variables X_1, X_2, X_3, X_4, and X_5. The constraint condition is the aquaculture area (Table 8.2). The benefit coefficient refers

Table 8.1 Effort allocation of various types of fishing boats in Shandong Province (Gao et al. 1999)

Fishing type		Total power (10^4 KW) and the number of fishing boats		
Trawler	Level	Above 441 KW	184–294 KW	136–147 KW
	Power (10^4 KW)	11	11	7.35
	Number of ships (boats)	250	500	500
Drift jigging boat	Level	Above 44.1 KW	29.4–44 KW	15 KW
	Power (10^4 KW)	11	3.68	7.35
	Number of ships (boats)	2500	1250	5000
Stake net	Level	15.4–44 KW	15 KW	8.8 KW
	Power (10^4 KW)	3.68	11	7.35
	Number of ships (boats)	850	7500	8330

Table 8.2 The constraint conditions unit: 10^4 mu (Gao et al. 1999)

Year	Fish farming	Shrimp and crab farming	Algae farming	Shallow sea shellfish farming	Mudflat shellfish farming	Shallow sea farming	Mudflat farming	Harbor farming
1994	1.32	65.98	17.66	73.67	37.66	92.40	52.82	52.13
2000	2.57	54.01	38.41	129.01	89.80	149.71	163.42	54.01

to the coefficient of each decision variable in the objective function. This study takes the net income per unit area by industry in 1994 as the reference benefit coefficient.

The linear programming model for the year 2000 is constructed as follows (Gao et al. 1999):

$$Z_{\max} = 1218X_1 + 179X_2 + 1243X_3 + 1341X_4 + 1361X_5$$

$$X_1 + X_2 + X_3 + X_4 + X_5 \leq 367.14$$

$$X_1 + X_2 + X_5 \leq 163.42 + 54.01$$

$$X_3 + X_4 \leq 149.71$$

$$X_1 \geq 2.57, X_2 \geq 54.01, X_3 \leq 38.41, X_4 \leq 129.01, X_5 \leq 89.80$$

Obtained by calculation

$$X_1 = 2.57, X_2 = 54.01, X_3 = 38.41, X_4 = 111.3, X_5 = 50.26$$

Total suitable farming area: $X_1 + X_2 + X_3 + X_4 + X_5 = 256.55$ (10^4 mu)

Through the analysis of the average yield of each subindustry from 1985 to 1994, the average yield of each subindustry in 2000 was calculated. According to the above linear programming model, the suitable aquaculture area value of each industry under the optimal benefit is obtained, and the expected output of the marine aquaculture industry in Shandong Province in 2000 is obtained:

1. The expected yield of mariculture in 2000 based on the subindustry average yield

$$0.55X_1 + 0.03X_2 + 2.20X_3 + 1.14X_4 + 0.58X_5$$
$$= 1.42 + 1.64 + 84.46 + 126.93 + 29.17$$
$$= 243.62\ \left(10^4\ \text{t}\right)$$

2. The expected yield of mariculture in 2000 based on the total average yield of the mariculture industry

$$0.87 \times 256.55 = 223.20\ \left(10^4\ \text{t}\right)$$

On the basis of a comprehensive investigation, study, and analysis of the current situation of marine fisheries in Shandong Province, this study used gray model theory to simulate and predict the characteristics and trends of fishery development in Shandong Province and used the theory and method of gray linear programming to calculate various grades of fishery development. The optimal estimation of the power of the fishing vessel and the optimal structure of fishing boats are calculated. Based on the evaluation results of this model, some suggestions for adjusting the structure of marine fisheries in Shandong Province are proposed (Gao et al. 1999).

References

Chen XJ (2003) Application of gray system theory in fishery science. China Agricultural Press. (In Chinese)

Chen XJ (2023) Application of gray system theory in fishery science. China Agricultural Press. (In Chinese)

Gao QL, Qiu TX, Song XF et al (1999) The study on the structure regulation of marine fishery of Shandong Province[J]. J Ocean Univ Qingdao 29(2):47–55. (In Chinese)

The manufacturer's authorised representative in the EU is Springer Nature Customer Service Centre GmbH, Europaplatz 3, 69115 Heidelberg, Germany. If you have any concerns regarding our products, please contact ProductSafety@springernature.com

Printed and bound by CPI Group (UK) Ltd, Croydon, CR0 4YY
01/10/2026
02225917-0003